PROF. MSDOSS MATHS SERIES II

TRIGONOMETRY
FORMULAE PRACTICE
WORKBOOK

By Prof. M. Subbiah Doss ©

Author's email Id: subdoss2014@gmail.com

No. of pages: 84

Year of publication 2016

> Author of
>
> PROF. MSDOSS MATHS BOOK SERIES I
>
> CALCULUS I
> DIFFERENTIAL CALCULUS
> FORMULAE PRACTICE
> WORKBOOK
>
> Prof. M. Subbiah Doss
>
> FOR THOSE STUDENTS WHO WANT TO HAVE A STRONG BASE IN DIFFERENTIAL CALCULUS FORMULAE

THAASU PUBLICATIONS,

14, II Cross Street,

Viswanathapuram,

Madurai – 625014,

Tamilnadu, India.

FOREWORD

Dear students,

This WORK BOOK in TRIGONOMETRY is suitable for those who want to remember the formulae in trigonometry forever. We all know that these formulae play a very important role in solving problems not only in trigonometry but in many other branches of mathematics. Thus it is mandatory for the students to remember all these formulae.

This workbook helps the students to have a thorough knowledge of the formulae in trigonometry and it helps the students to apply the appropriate formula in the appropriate place.

Students are advised to follow the following steps:

1. If you close your eyes, you can see a screen. Can't you?

Whenever you find the symbol ☺ in this workbook, close your eyes and try to recollect the given formulae on that screen. Try to recollect the formulae on the screen for at least 3 times.

Do you ever find it difficult to remember and recollect the names of your family members, friends and close relatives? Just like that you can remember and recollect the formulae without any pain.

2. Immediately after recollecting the formulae, you are advised to apply these formulae in solving the problems given in the workbook. Try to solve the exercises as many times as possible (at proper interval)

3. After this practice, try to solve the problems given in your text book. No need to say that you will find it easy to solve the problems.

4. Question: I want to be a master in trigonometry. What should I do?

   Answer: Well, it is a nice question indeed.

   To become a master in trigonometry, try to solve the problems given in the book,

   '**PLANE TRIGONOMETRY**' (Part I), S.L.LONEY

Also read this: "the mathematical genius Srinivasa Ramanujan won prizes for his outstanding performance in mathematics and mastered **Loney's Trigonometry** in his fourth year at school" – 'SRINIVASA RAMANUJAN - A mathematical genius' – K. Srinivasa Rao.

■ Aim of this **'formulae practice workbook:**

To help the students to **remember, recollect and apply** the various formulae in the appropriate place while solving the problems.

■ Already tested in India among average and below average higher secondary students (11th & 12th std) with very good results.

■ *Theory is not discussed here in detail.*

■ More number of solved problems and problems for practice with solutions.

■ A self evaluation test with answers.

■ Practice! Practice!

This practice helps you

• to discard your pre-conceived ideas of Trigonometry. You can be friendly (familiar) with the '*truck load of formulae*' which is frightening you so far.

• to solve the problems given in the text book and assignment sheets easily and independently.

• to understand the theory given in your text book without any fear.

You can do it! No doubt!!

■ The thorough knowledge acquired here will be more useful not only in Differential Calculus but also in Integral Calculus, Differential Equations etc.

■ This workbook is available at Amazon.com

Wish you all the best!

Prof. M. Subbiah Doss

CONTENTS

Unit – 1	Trigonometrical ratios	01
Unit – 2	Trigonometrical ratios of	
	i. Particular angles	10
	ii. Related angles	12
Unit – 3	Addition formulae	21
	Multiple angles	25
Unit – 4	Product formulae	31
Unit – 5	Trigonometrical equations	36
Unit – 6	Inverse trigonometric functions	41
Unit – 7	Properties of triangle	47
Appendix I	How do the ratios look like?	61
Appendix II	Measurements of angles	65
Appendix III	Values of the trigonometrical ratios in some special cases	66
Self Evaluation Test		67
Tables		70
Practice! Practice!		74

UNIT– 1

1.1 Trigonometrical Ratios

Let O be a fixed point and OX be the initial fixed line.

Let A be any point on OX.

Rotate OA through an angle θ in the anti-clockwise direction. Let the new position be OP

Draw PM perpendicular to OX. Then angle AOP = θ ($< 90^0$)

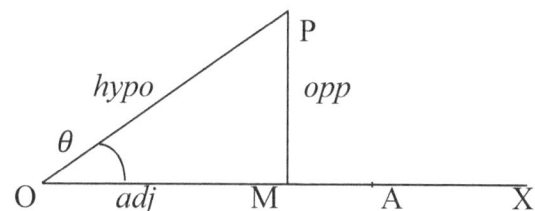

In the right angled triangle OMP, the side OP is the hypotenuse, MP, the opposite side and OM, the adjacent side (base).

Here we define three trigonometrical ratios as follows.

Sine of the angle θ ie. $\sin \theta = \dfrac{opp}{hypo} = \dfrac{MP}{OP}$

Cosine of the angle θ ie. $\cos \theta = \dfrac{adj}{hypo} = \dfrac{OM}{OP}$

Tangent of the angle θ ie $\tan \theta = \dfrac{opp}{adj} = \dfrac{MP}{OM}$

The following three trigonometrical ratios are the *reciprocal* of the above three ratios.

Cosecant of the angle θ ie. $\csc \theta = \dfrac{hypo}{opp} = \dfrac{OP}{MP} = \dfrac{1}{\sin \theta}$

Secant of the angle θ ie. $\sec \theta = \dfrac{hypo}{adj} = \dfrac{OP}{OM} = \dfrac{1}{\cos \theta}$

Co-tangent of the angle θ ie $\cot \theta = \dfrac{adj}{opp} = \dfrac{OM}{MP} = \dfrac{1}{\tan \theta}$

Thus we have

$\sin \theta$	$\csc \theta$
$\cos \theta$	$\sec \theta$
$\tan \theta$	$\cot \theta$

Relation between the above six ratios

$\frac{1}{\sin\theta} = \text{cosec}\,\theta$	$\frac{1}{\text{cosec}\,\theta} = \sin\theta$	$\tan\theta = \frac{\sin\theta}{\cos\theta}$
$\frac{1}{\cos\theta} = \sec\theta$	$\frac{1}{\sec\theta} = \cos\theta$	$\cot\theta = \frac{\cos\theta}{\sin\theta}$
$\frac{1}{\tan\theta} = \cot\theta$	$\frac{1}{\cot\theta} = \tan\theta$	

Examples 1.1

1. $\cos\theta \cdot \sec\theta = \cos\theta \cdot \frac{1}{\cos\theta} = 1$

2. $\tan\theta \cdot \cos\theta = \frac{\sin\theta}{\cos\theta} \cdot \cos\theta = \sin\theta$

3. $\frac{\sin\theta}{\tan\theta} = \sin\theta \cdot \cot\theta = \sin\theta \cdot \frac{\cos\theta}{\sin\theta} = \cos\theta$

4. $\frac{\cot\theta}{\cos\theta} - \text{cosec}\,\theta = \frac{\cos\theta}{\sin\theta} \cdot \frac{1}{\cos\theta} - \frac{1}{\sin\theta} = \frac{1}{\sin\theta} - \frac{1}{\sin\theta} = 0$

5. $\frac{\cos\theta}{\sin\theta} \cdot \frac{1}{\text{cosec}\,\theta} - \frac{1}{\sec\theta} = \frac{\cos\theta}{\sin\theta} \cdot \sin\theta - \cos\theta = 0$

Exercise 1.1 : Use the above results to solve the following

1. $\sin\theta \cdot \text{cosec}\,\theta = $ ....

   a) 0 b) 1 c) -1 d) $\tan\theta$

2. $\cot\theta \cdot \sin\theta = $ ....

   a) $\cos\theta$ b) 1 c) $\cos\theta$ d) $\cot\theta$

3. $\frac{\cos\theta}{\cot\theta} = $ ....

   a) 1 b) $\cos\theta$ c) $\sin\theta$ d) $\tan\theta$

4. $\sec\theta \cdot \cos\theta + \text{cosec}\,\theta \cdot \sin\theta - \tan\theta \cdot \cot\theta = $ ....

   a) 1 b) 0 c) -1 d) 2

5. $\frac{\tan\theta}{\sin\theta} - \sec\theta = $ ....

a) 1 b) 0 c) 2 sec θ d) − 1

Answers : 1 – b 2 – c 3 – c 4 – a 5 – b

Solution 1.1

1. $\sin\theta \cdot \operatorname{cosec}\theta = \sin\theta \cdot \dfrac{1}{\sin\theta} = 1$

2. $\cot\theta \cdot \sin\theta = \dfrac{\cos\theta}{\sin\theta} \cdot \sin\theta = \cos\theta$

3. $\dfrac{\cos\theta}{\cot\theta} = \cos\theta \cdot \tan\theta = \cos\theta \cdot \dfrac{\sin\theta}{\cos\theta} = \sin\theta$

4. $\sec\theta \cdot \cos\theta + \operatorname{cosec}\theta \cdot \sin\theta - \tan\theta \cdot \cot\theta$

   $= \dfrac{1}{\cos\theta} \cdot \cos\theta + \dfrac{1}{\sin\theta} \cdot \sin\theta - \tan\theta \cdot \dfrac{1}{\tan\theta} = 1 + 1 - 1 = 1$

5. $\dfrac{\tan\theta}{\sin\theta} - \sec\theta = \dfrac{\sin\theta}{\cos\theta} \cdot \dfrac{1}{\sin\theta} - \sec\theta = \sec\theta - \sec\theta = 0$

Whenever you find the symbol ☺, close your eyes and try to recollect the given formulae on the screen. Try to recollect the formulae on the screen for at least 3 times. Try! You can do it!

Table 1.1 ☺ **Try to recollect the following**

sin θ	cosec θ	
cos θ	sec θ	
tan θ	cot θ	
$\dfrac{1}{\sin\theta} = \operatorname{cosec}\theta$ $\dfrac{1}{\cos\theta} = \sec\theta$ $\dfrac{1}{\tan\theta} = \cot\theta$	$\dfrac{1}{\operatorname{cosec}\theta} = \sin\theta$ $\dfrac{1}{\sec\theta} = \cos\theta$ $\dfrac{1}{\cot\theta} = \tan\theta$	$\tan\theta = \dfrac{\sin\theta}{\cos\theta}$ $\cot\theta = \dfrac{\cos\theta}{\sin\theta}$

1.2 Relation between the above six ratios.

In the right angled triangle OMP,

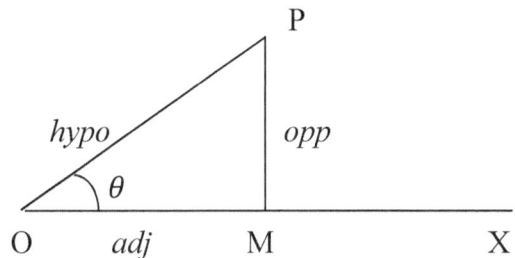

By Pythagorean theorem, we have

$$MP^2 + OM^2 = OP^2$$

i) Dividing by OP^2, we get $\dfrac{MP^2}{OP^2} + \dfrac{OM^2}{OP^2} = 1$

$$\Rightarrow \quad sin^2\theta + cos^2\theta = 1$$

ii) Dividing by OM^2, we get $\dfrac{MP^2}{OM^2} + 1 = \dfrac{OP^2}{OM^2}$

$$\Rightarrow \quad tan^2\theta + 1 = sec^2\theta$$

iii) Dividing by MP^2, we get $1 + \dfrac{OM^2}{MP^2} = \dfrac{OP^2}{MP^2}$

$$\Rightarrow \quad 1 + cot^2\theta = cosec^2\theta$$

 ☺ **Try to recollect the following**

$\sin^2\theta + \cos^2\theta = 1$	
$\sin^2\theta = 1 - \cos^2\theta$;	$\cos^2\theta = 1 - \sin^2\theta$
$1 + \tan^2\theta = \sec^2\theta$	$1 + \cot^2\theta = \csc^2\theta$
$\sec^2\theta - 1 = \tan^2\theta$	$\csc^2\theta - 1 = \cot^2\theta$

Examples 1.2

1. $\tan\theta . \cot\theta - \dfrac{\sin\theta}{\csc\theta} = 1 - \sin^2\theta = \cos^2\theta$

2. $\sin\theta . \csc\theta - \dfrac{\cos\theta}{\sec\theta} = 1 - \cos^2\theta = \sin^2\theta$

3. $\dfrac{1}{\cos^2\theta} - 1 = \sec^2\theta - 1 = \tan^2\theta$

4. $\sqrt{\dfrac{\text{cosec }\theta}{\sin \theta} - 1} \quad = \sqrt{\text{cosec}^2\,\theta - 1} \quad = \sqrt{\cot^2 \theta} = \cot \theta$

5. $1 - \dfrac{\tan^2 \theta}{\sec^2 \theta} = 1 - \dfrac{\sin^2 \theta}{\cos^2 \theta} \cdot \cos^2\theta = 1 - \sin^2 \theta = \cos^2 \theta$

Fill in the blanks

6. $\cot \theta \times (\ldots\ldots) - \cos \theta = 0$

7. $\dfrac{1}{\sin^2\theta} \ldots 1 = \cot^2 \theta \qquad [\ + \ (or) \ - \ (or) \ \times\ ]$

8. $\ldots\ldots \times \sec \theta + \tan \theta \times \ldots\ldots = 2$

Answers : 6. $\sin \theta$ 7. $-$ve sign 8. $\cos \theta$, $\cot \theta$

Complete the missing steps

9. $\tan^4 \theta - \sec^4 \theta = \ldots\ldots\ldots\ldots\ldots\ldots\ldots$

$\qquad\qquad\qquad\qquad = \ldots\ldots\ldots\ldots\ldots\ldots\ldots$

$\qquad\qquad\qquad\qquad = -(\tan^2 \theta + \sec^2 \theta)$

Solution

$\tan^4 \theta - \sec^4 \theta \quad = (\tan^2 \theta + \sec^2 \theta)(\tan^2 \theta - \sec^2 \theta)$

$\qquad\qquad\qquad\quad = -(\sec^2 \theta - \tan^2 \theta)(\tan^2 \theta + \sec^2 \theta)$

$\qquad\qquad\qquad\quad = -(\tan^2 \theta + \sec^2 \theta) \quad \text{since} \ \sec^2 \theta - \tan^2 \theta = 1$

10. $\sin^2 \theta - \sin^4 \theta \ = \ \ldots\ldots\ldots\ldots\ldots\ldots$

$\qquad\qquad\qquad\quad = \ \ldots\ldots\ldots\ldots\ldots\ldots$

$\qquad\qquad\qquad\quad = \cos^2 \theta - \cos^4 \theta$

Solution

$\sin^2 \theta - \sin^4 \theta \ = \sin^2 \theta\,(1 - \sin^2 \theta)$

$\qquad\qquad\qquad\quad = (1 - \cos^2 \theta)\cos^2 \theta$

$\qquad\qquad\qquad\quad = \cos^2 \theta - \cos^4 \theta$

Exercise 1.2

A : Choose the correct answer

1. $\dfrac{\sin\theta}{\text{cosec}\,\theta} + \dfrac{\cos\theta}{\sec\theta} = $ ....

   a) 0   b) 1   c) –1   d) $2\tan\theta$

2. $1 + \dfrac{\tan\theta}{\cot\theta} = $ ....

   a) $\sec^2\theta$ b) $\sec\theta$ c) 1   d) 0

3. $\dfrac{1}{\sin^2\theta} - 1 = $ ....

   a) $\tan^2\theta$ b) $\sin^2\theta$ c) $\cot^2\theta$ d) $-\cot^2\theta$

4. $\dfrac{\cos\theta}{\sec\theta} + \dfrac{\cot\theta}{\tan\theta} + \dfrac{\sin\theta}{\text{cosec}\,\theta} = $ ....

   a) $\sec^2\theta$ b) $\cot^2\theta$ c) $\tan^2\theta$ d) $\text{cosec}^2\theta$

5. $\sqrt{\dfrac{\sec\theta}{\cos\theta} - 1} = $ ....

   a) $\sin\theta$ b) $\tan\theta$ c) $\cot\theta$ d) $\cos\theta$

6. $1 - \dfrac{\cot^2\theta}{\text{cosec}^2\theta} = $ ....

   a) $\sin^2\theta$ b) $\cos^2\theta$ c) $\tan^2\theta$ d) $\sec^2\theta$

7. $\dfrac{\sec\theta}{\sqrt{1-\sin^2\theta}} - \dfrac{\sin^2\theta}{1-\sin^2\theta} = $ ....

   a) 0   b) 1   c) –1   d) $\sin\theta$

8. $\dfrac{1}{\sin^2\theta} - (1 - \sin^2\theta)\,\text{cosec}^2\theta = $ ....

   a) $\cos^2\theta$ b) $\sin^2\theta$ c) 1   d) 0

9. $\tan\theta \cdot \cot\theta - \text{cosec}\,\theta \cdot \sin\theta = $ ....

   a) 1   b) –1   c) 0   d) $2\sin\theta$

10. $\sin^2\theta \cdot \text{cosec}^2\theta + \cos^2\theta \cdot \sec^2\theta = $ ....

    a) 1   b) 2   c) –1   d) 0

B Complete the following

1. $\dfrac{\sin\theta}{\text{......}} + \dfrac{\cos\theta}{\sec\theta} = 1$

2. $\text{......} + \dfrac{\cot\theta}{\tan\theta} = \text{cosec}^2\theta$

3. $\dfrac{\cos\theta}{\sec\theta} + \dfrac{\cot\theta}{\text{......}} + \dfrac{\sin\theta}{\text{cosec}\,\theta} = \text{cosec}^2\theta$

4. $\dfrac{1}{\text{......}} - 1 = \tan^2\theta$

5. $\sqrt{\dfrac{\text{cosec}\,\theta}{\sin\theta} - 1} = \text{.........}$

6. $1 - \dfrac{\text{......}}{\sec^2\theta} = \cos^2\theta$

7. $\dfrac{\cos^2\theta}{\sin^2\theta} \times \text{.........} + \text{..........} \times \tan^2\theta = 2$

8. $(1 - \cos^2\theta) \times \dfrac{1}{\text{......}} = \sin^4\theta$

9. $\dfrac{\text{......}}{\sqrt{\text{cosec}^2\theta - 1}} = \sec\theta$

10. $\dfrac{\sqrt{1 + \tan^2\theta}}{\text{......}} = \text{cosec}\,\theta$

C Complete the missing steps.

1. $\text{cosec}^4\theta - \text{cosec}^2\theta = \text{cosec}^2\theta\,(\text{cosec}^2\theta - 1)$
   $= \text{................}$
   $= \cot^2\theta + \cot^4\theta$

2. $\cos^4 x - \sin^4 x = \text{......................}$
   $= \cos^2 x - \sin^2 x$
   $= \text{......................}$
   $= 1 - 2\sin^2 x$

Answers :

A 1 – b 2 – a 3 – c 4 – d 5 – b 6 – a 7 – b 8 – c 9 – c 10 – b

B: 1. cosec θ   2. 1   3. tan θ   4. $\cos^2 θ$ 5. cot θ   6. $\tan^2 θ$

   7. $\tan^2 θ$, $\cot^2 θ$ 8. $\csc^2 θ$ 9. cosec θ   10. tan θ

C : 1. $(1 + \cot^2 θ) \cot^2 θ$

    2. $(\cos^2 x - \sin^2 x)(\cos^2 x + \sin^2 x)$, $1 - \sin^2 x - \sin^2 x$

Solution 1.2 A (Here only 1.2 A is provided with solution)

1. $\dfrac{\sin θ}{\csc θ} + \dfrac{\cos θ}{\sec θ} = \sin θ \cdot \sin θ + \cos θ \cdot \cos θ$

$$= \sin^2 θ + \cos^2 θ = 1$$

2. $1 + \dfrac{\tan θ}{\cot θ} = 1 + \tan θ \cdot \tan θ = 1 + \tan^2 θ = \sec^2 θ$

3. $\dfrac{1}{\sin^2 θ} - 1 = \csc^2 θ - 1 = \cot^2 θ$

4. $\dfrac{\cos θ}{\sec θ} + \dfrac{\cot θ}{\tan θ} + \dfrac{\sin θ}{\csc θ} = \cos^2 θ + \cot^2 θ + \sin^2 θ$

$$= 1 + \cot^2 θ = \csc^2 θ$$

5. $\sqrt{\dfrac{\sec θ}{\cos θ} - 1} = \sqrt{\sec^2 θ - 1} = \sqrt{\tan^2 θ} = \tan θ$

6. $1 - \dfrac{\cot^2 θ}{\csc^2 θ} = \dfrac{\csc^2 θ - \cot^2 θ}{\csc^2 θ}$

$$= \dfrac{1}{\csc^2 θ} = \sin^2 θ$$

7. $\dfrac{\sec θ}{\sqrt{1-\sin^2 θ}} - \dfrac{\sin^2 θ}{1-\sin^2 θ} = \dfrac{\sec θ}{\cos θ} - \dfrac{\sin^2 θ}{\cos^2 θ}$

$$= \sec^2 θ - \tan^2 θ = 1$$

8. $\dfrac{1}{\sin^2\theta} - (1 - \sin^2\theta) \cdot \text{cosec}^2\theta$

$$= \text{cosec}^2\theta - (1 - \sin^2\theta) \cdot \text{cosec}^2\theta$$

$$= \text{cosec}^2\theta \, (1 - 1 + \sin^2\theta)$$

$$= \text{cosec}^2\theta \cdot \sin^2\theta = 1$$

9. $\tan\theta \cdot \cot\theta - \text{cosec}\,\theta \cdot \sin\theta = \tan\theta \cdot \dfrac{1}{\tan\theta} - \dfrac{1}{\sin\theta} \cdot \sin\theta = 0$

10. $\sin^2\theta \cdot \text{cosec}^2\theta + \cos^2\theta \cdot \sec^2\theta$

$$= \sin^2\theta \cdot \dfrac{1}{\sin^2\theta} + \cos^2\theta \cdot \dfrac{1}{\cos^2\theta} = 1 + 1 = 2$$

Table 1.2

☺ **Try to recollect the following**

sin θ	cosec θ
cos θ	sec θ
tan θ	cot θ

| $\dfrac{1}{\sin\theta} = \text{cosec}\,\theta$ $\dfrac{1}{\cos\theta} = \sec\theta$ $\dfrac{1}{\tan\theta} = \cot\theta$ | $\dfrac{1}{\text{cosec}\,\theta} = \sin\theta$ $\dfrac{1}{\sec\theta} = \cos\theta$ $\dfrac{1}{\cot\theta} = \tan\theta$ | $\tan\theta = \dfrac{\sin\theta}{\cos\theta}$ $\cot\theta = \dfrac{\cos\theta}{\sin\theta}$ |

$\sin^2\theta + \cos^2\theta = 1$

$\sin^2\theta = 1 - \cos^2\theta \quad ; \quad \cos^2\theta = 1 - \sin^2\theta$

| $1 + \tan^2\theta = \sec^2\theta$ $\sec^2\theta - 1 = \tan^2\theta$ | $1 + \tan^2\theta = \sec^2\theta$ $\sec^2\theta - 1 = \tan^2\theta$ |

UNIT 2

2.1 Trigonometrical ratios of particular angles

How to remember the values?

Write down the particular angles in the ascending order

	0°	30°	45°	60°	90°
in the top row →					
Write down →	0	1	2	3	4
Divide by 4 →	$\frac{0}{4}$	$\frac{1}{4}$	$\frac{2}{4}$	$\frac{3}{4}$	$\frac{4}{4}$
Take square root →	$\sqrt{\frac{0}{4}}$	$\sqrt{\frac{1}{4}}$	$\sqrt{\frac{1}{2}}$	$\sqrt{\frac{3}{4}}$	$\sqrt{\frac{4}{4}}$
$\sin\theta$ →	0	$\frac{1}{2}$	$\frac{1}{\sqrt{2}}$	$\frac{\sqrt{3}}{2}$	1
Write the above values in the reverse order $\cos\theta$ →	1	$\frac{\sqrt{3}}{2}$	$\frac{1}{\sqrt{2}}$	$\frac{1}{2}$	0
Divide $\sin\theta$ by $\cos\theta$, we get $\tan\theta$ →		$\frac{1}{\sqrt{3}}$			∞
$\cot\theta$ →				$\frac{1}{\sqrt{3}}$	
$\sec\theta$ →		$\frac{2}{\sqrt{3}}$			
$\mathrm{cosec}\,\theta$ →				$\frac{2}{\sqrt{3}}$	

(**Complete the table**)

Examples 2.1

1. $\sin^2 30 + \sin^2 45 + \sin^2 60 \;=\; \left(\frac{1}{2}\right)^2 + \left(\frac{1}{\sqrt{2}}\right)^2 + \left(\frac{\sqrt{3}}{2}\right)^2$

$$= \frac{1}{4} + \frac{1}{2} + \frac{3}{4} = \frac{3}{2}$$

2. $\sin 45 . \sin 60 - \cos 45 . \cos 60 = \frac{1}{\sqrt{2}} . \frac{\sqrt{3}}{2} - \frac{1}{\sqrt{2}} . \frac{1}{2} = \frac{\sqrt{3}-1}{2\sqrt{2}}$

3. If $A = 60°$, verify that $\tan^2 A = \sec^2 A - 1$

    L.H.S $= \tan^2 60 = (\sqrt{3})^2 = 3$

    R.H.S $= \sec^2 60 - 1 = 4 - 1 = 3$

4. If $A = 45°$, verify that $\tan 2A = \frac{2\tan A}{1-\tan^2 A}$

    L.H.S $= \tan 90 = \infty$

    R.H.S $= \frac{2\tan 45}{1-\tan^2 45} = \frac{2}{0} = \infty$

5. $\sin \frac{\pi}{6} . \cos \frac{\pi}{3} + \cos \frac{\pi}{6} . \sin \frac{\pi}{3} = \sin 30 . \cos 60 + \cos 30 . \sin 60$

$$= \frac{1}{2} . \frac{1}{2} + \frac{\sqrt{3}}{2} . \frac{\sqrt{3}}{2} = \frac{1}{4} + \frac{3}{4} = 1$$

Exercise 2.1 Choose the correct answer

1. $\cos 45° =$ ....
a) $\tan 45°$ b) $\sin 45°$ c) $\sec 45°$ d) $\operatorname{cosec} 45°$

2. $\sin 0° =$ ....
a) $\cot 0°$ b) $\cos 0°$ c) $\cos 90°$ d) $\tan 90°$

3. $\operatorname{cosec} 90° =$ ....
a) $\sec 90°$ b) $\cot 90°$ c) $\cos 90°$ d) $\sin 90°$

4. $\sec 45° =$ ....
a) $\tan 45°$ b) $\sin 60°$ c) $\operatorname{cosec} 45°$ d) $\cos 60°$

5. $\sin^2 \frac{\pi}{2} + \cos^2 \frac{\pi}{2} =$ ....
a) 0 b) 1 c) 2 d) ∞

6. $\cos 90° + 4\cos^2 45° \sin 30° \tan 45° =$ ....

a) 0 b) 2 c) 1 d) $\frac{1}{2}$

7. If $0 < x° < 90°$, the numerical value of x for which $\sin x° = \cos x°$ is ....
a) 0° b) 30° c) 45° d) 60°

8. The ratio ($\sin 30° + \cos 60°$) to $\cot 45°$ is ....
a) 1 : 2 b) 2 : 1 c) 1 : 1 d) 3 : 1

9. $(\frac{1}{\cos 45°} - \sin 45°)^2 (\tan 45° + \frac{1}{\cot 45°}) = $ ....
a) 1 b) 2 c) 0 d) $\frac{1}{3}$

10. If $\theta < 90°$ and $4\sin^2\theta - 1 = 0$, then the value of θ is ....
a) 0° b) 30° c) 45° d) 60°

Answers : (only answers are given)

1 – b   2 – c   3 – d   4 – c   5 – b   6 – c   7 – c   8 – c   9 – a   10 – b

2.2 Trigonometrical ratios of related angles

2.2.1 A **shortcut method** to find trigonometrical ratios of related angles

Fig 1   Set I

II quadrant	90° I quadrant
sin θ + ve	All ratios + ve
cosec θ + ve	
180°	0°, 360°
tan θ + ve	cos θ + ve
cot θ + ve	sec θ + ve
III quadrant	270° IV quadrant

Set I:
- $-\theta$
- $180 - \theta$
- $180 + \theta$
- $360 - \theta$
- $360 + \theta$

Remember the following rules

i. For the angles in the **set I**, there is no change in the ratio.

ii. **the sign of the given ratio** (+ ve or – ve) is decided by Fig 1.

Examples

1. Consider the ratio $\sin(180 + \theta)$. Here,

   i. $180 + \theta$ lies in the **set I** (see the box above)

      $\Rightarrow$ by (i), there is no change in ratio. we get $\sin\theta$ only

   ii. $180 + \theta$ lies in the **III quadrant**

      $\Rightarrow$ by (ii), the sign of the given ratio is $-$ve

      Hence $\sin(180 + \theta) = -\sin\theta$

2. Consider the ratio $\sec(360 - \theta)$. Here,

   i. $360 - \theta$ lies in the **set I** (see the box above)

      $\Rightarrow$ by (i), there is no change in ratio. we get $\sec\theta$ only

   ii. $360 - \theta$ lies in the **IV quadrant**

      $\Rightarrow$ by (ii), the sign of the given ratio is $+$ve

      Hence $\sec(360 - \theta) = \sec\theta$

3. Consider the ratio $\tan(-\theta)$. Here,

   i. $-\theta$ lies in the **set I** (see the box above)

      $\Rightarrow$ by (i), there is no change in ratio. we get $\tan\theta$ only

   ii. $-\theta$ lies in the **IV quadrant**

      $\Rightarrow$ by (ii), the sign of the given ratio is $-$ve

      Hence $\tan(-\theta) = -\tan\theta$

Exercise 2.2.1

Complete the following. Compare your answers with the table 5 given towards the end of this workbook.

| $-\theta$ | $180-\theta$ | $180+\theta$ |

sin (−θ) = ...   sin (180 − θ) = ...   sin (180 + θ) = ...

cos (−θ) = ...   cos (180 − θ) = ...   cos (180 + θ) = ...

tan (−θ) = ...   tan (180 − θ) = ...   tan (180 + θ) = ...

cosec (−θ) = ...   cosec (180 − θ) = ...   cosec (180 + θ) = ...

sec (−θ) = ...   sec (180 − θ) = ...   sec (180 + θ) = ...

cot (−θ) = ...   cot (180 − θ) = ...   cot (180 + θ) = ...

| $360-\theta$ | $360+\theta$ |

sin (360 − θ) = ...   sin (360 + θ) = ...

cos (360 − θ) = ...   cos (360 + θ) = ...

tan (360 − θ) = ...   tan (360 + θ) = ...

cosec (360 − θ) = ...   cosec (360 + θ) = ...

sec (360 − θ) = ...   sec (360 + θ) = ...

cot (360 − θ) = ...   cot (360 + θ) = ...

2.2.2 shortcut method to find trigonometrical ratios of related angles

Fig 1

Set II

II quadrant	90°	I quadrant
sin θ +ve		All ratios +ve
cosec θ +ve		
180°		0°, 360°
tan θ +ve		cos θ +ve
cot θ +ve		sec θ +ve
III quadrant	270°	IV quadrant

| $90-\theta$ |
| $90+\theta$ |
| $270-\theta$ |
| $270+\theta$ |

Remember the following rules

i. For the angles in the **set II**, there is a change in the ratio.

$$\sin \Leftrightarrow \cos \;;\quad \tan \Leftrightarrow \cot \;;\quad \csc \Leftrightarrow \sec$$

ii. **the sign of the given ratio** (+ ve or – ve) is decided by Fig 1.

Examples

1 Consider the ratio $\sin(90 + \theta)$. Here,

i. **90 + θ** lies in the **set II** (see the box above)

   => by (i), there is change in ratio. we get **cos θ**

ii. **90 + θ** lies in the **II quadrant**

   => by (ii), the sign of the *given ratio* **sin** is + ve

   Hence $\sin(90 + \theta) = +\cos\theta$

2 consider the ratio $\sec(270 + \theta)$. Here,

i. **270 + θ** lies in the **set II** (see the box above)

   => by (i), there is change in ratio. we get **cosec θ**

ii. **270 + θ** lies in the **IV quadrant**

   => by (ii), the sign of the *given ratio* **sec** is + ve

   Hence $\sec(270 + \theta) = +\csc\theta$

3 Consider the ratio $\cos(270 - \theta)$. Here,

i. **270 − θ** lies in the **set II** (see the box above)

   => by (i), there is change in ratio. we get **sin θ**

ii. **270 − θ** lies in the **III quadrant**

   => by (ii), the sign of the *given ratio* **cos** is − ve

   Hence $\cos(270 - \theta) = -\sin\theta$

Exercise 2.2.2 Complete the following. Compare your answers with the table 5 given towards the end of this workbook.

$90 - \theta$ $\qquad$ $90 + \theta$ $\qquad$ $270 - \theta$

$\sin(90 - \theta) = \ldots\ldots$ $\quad$ $\sin(90 + \theta) = \ldots\ldots$ $\quad$ $\sin(270 - \theta) = \ldots\ldots$

$\cos(90 - \theta) = \ldots\ldots$ $\quad$ $\cos(90 + \theta) = \ldots\ldots$ $\quad$ $\cos(270 - \theta) = \ldots\ldots$

$\tan(90 - \theta) = \ldots\ldots$ $\quad$ $\tan(90 + \theta) = \ldots\ldots$ $\quad$ $\tan(270 - \theta) = \ldots\ldots$

$\operatorname{cosec}(90 - \theta) = \ldots\ldots$ $\quad$ $\operatorname{cosec}(90 + \theta) = \ldots\ldots$ $\quad$ $\operatorname{cosec}(270 - \theta) = \ldots\ldots$

$\sec(90 - \theta) = \ldots\ldots$ $\quad$ $\sec(90 + \theta) = \ldots\ldots$ $\quad$ $\sec(270 - \theta) = \ldots\ldots$

$\cot(90 - \theta) = \ldots\ldots$ $\quad$ $\cot(90 + \theta) = \ldots\ldots$ $\quad$ $\cot(270 - \theta) = \ldots\ldots$

$270 + \theta$

$\sin(270 + \theta) = \ldots\ldots$ $\quad$ $\cos(270 + \theta) = \ldots\ldots$ $\quad$ $\tan(270 + \theta) = \ldots\ldots$

$\operatorname{cosec}(270 + \theta) = \ldots\ldots$ $\quad$ $\sec(270 + \theta) = \ldots\ldots$ $\quad$ $\cot(270 + \theta) = \ldots\ldots$

Table 2.2

☺ **Try to recollect the following**

```
                             Fig 1
   |                                                    |
   |      II quadrant         90°     I quadrant        |
   |                           |                        |
   |        sin θ  + ve        |    All ratios  + ve    |
   |        cosec θ  + ve      |                        |
   |  180° _____|_____ 0°, 360°   |
   |                           |                        |
   |        tan θ  + ve        |      cos θ  + ve       |
   |        cot θ  + ve        |      sec θ  + ve       |
   |      III quadrant        270°    IV quadrant       |
```

> **Set I**
>
> $$-\theta, 180-\theta, 180+\theta, 360-\theta, 360+\theta$$
>
> **Remember the following rules**
>
> i. For the angles in the set **I**, **there is no change in the ratio.**
>
> ii. the sign of the given ratio (+ ve **or** – ve) is decided by fig 1.
>
> **Set II**
>
> $$90-\theta, \ 90+\theta, \ 270-\theta, \ 270+\theta$$
>
> **Remember the following rules**
>
> i. For the angles in the set **II**, **there is a change in the ratio.**
>
>   **sin $\Longleftrightarrow$ cos ;   tan $\Longleftrightarrow$ cot ;   cosec $\Longleftrightarrow$ sec**
>
> ii. the sign of the given ratio (+ ve **or** – ve) is decided by fig 1.

Exercise 2.2.3

Complete the following. Compare your answers with the table 5 given towards the end of this workbook.

$\tan(180-\theta) = \ldots\ldots$ $\sin(-\theta) = \ldots\ldots$

$\cos(270+\theta) = \ldots\ldots$ $\sec(90+\theta) = \ldots\ldots$

$\sin(90-\theta) = \ldots\ldots$ $\csc(180+\theta) = \ldots\ldots$

$\cot(360-\theta) = \ldots\ldots$ $\cot(90-\theta) = \ldots\ldots$

$\cos(90-\theta) = \ldots\ldots$ $\sec(90-\theta) = \ldots\ldots$

$\tan(-\theta) = \ldots\ldots$ $\cot(180-\theta) = \ldots\ldots$

$\sin(270-\theta) = \ldots\ldots$ $\cos(90+\theta) = \ldots\ldots$

$\cos(360+\theta) = \ldots\ldots$ $\sec(360+\theta) = \ldots\ldots$

2.2.4 Odd and even functions

1. Since $\sin(-\theta) = -\sin\theta$, $\sin\theta$ is an **odd** function

   and $\cos(-\theta) = \cos\theta$, $\cos\theta$ is an **even** function.

2. If $\varphi = n \times 360 + \theta$, where $n = 1, 2, 3,\ldots$ then $\varphi = \theta$

Examples 2.2.4

1. $\sin 1575° = \sin(4 \times 360 + 135°)$

   $= \sin 135° = \sin(90 + 45°) = \cos 45° = \dfrac{1}{\sqrt{2}}$

2. $\tan 1125° = \tan(3 \times 360 + 45°) = \tan 45° = 1$

3. $\operatorname{cosec}(-1500°) = -\operatorname{cosec} 1500°$

   $= -\operatorname{cosec}(4 \times 360 + 60°)$

   $= -\operatorname{cosec} 60° = -\dfrac{2}{\sqrt{3}}$

4. $\sec(-980°) = \sec 980° = \sec(2 \times 360° + 260°)$

   $= \sec 260° = \sec(270° - 10°) = -\operatorname{cosec} 10°$

5. $\dfrac{\cot(90-\theta)\,\sin(180+\theta)\,\sec(360-\theta)}{\tan(180+\theta)\,\sec(-\theta)\,\cos(90+\theta)}$

   $= \dfrac{\tan\theta \cdot (-\sin\theta) \cdot \sec\theta}{\tan\theta \cdot \sec\theta \cdot (-\sin\theta)} = 1$

Exercise 2.2.4   Choose the correct answer

1. If $\cos(-60) = \sin x$, then $x = \ldots\ldots$
   a) $90°$ b) $45°$ c) $30°$ d) $60°$

2. $\sin 610° = \ldots\ldots$
   a) $\sin 70°$ b) $-\cos 20°$ c) $\cos 70°$ d) $-\cos 70°$

3. $\cos 150 = \ldots\ldots$
   a) $-\dfrac{\sqrt{3}}{2}$ b) $\dfrac{\sqrt{3}}{2}$ c) $\dfrac{1}{2}$ d) $-\dfrac{1}{2}$

4. tan 1830 = ...

a) $\frac{1}{2}$ b) $-\frac{1}{2}$ c) $\frac{1}{\sqrt{3}}$ d) $-\frac{1}{\sqrt{3}}$

5. $\sin^2 52 + \sin^2 38$ = ...

a) 0 b) 1 c) −1 d) $\frac{1}{2}$

6. sin 36 sec 54 − cos 24 cosec 66 = ...

a) 0 b) 1 c) 2 d) −1

7. $\dfrac{\sin(90-A)}{\csc(90-A)} + \dfrac{\cos(90-A)}{\sec(90-A)} = $ ....

a) 0 b) 1 c) 2 d) −1

8. $1 - \dfrac{\sin(90-A)\sin A}{\tan A} = $ ....

a) $\cos^2 A$ b) $-\cos^2 A$ c) $-\sin^2 A$ d) $\sin^2 A$

9. cot (− 675) = ...

a) $\frac{1}{2}$ b) $-\frac{1}{2}$ c) 1 d) −1

Answers : 1 - c 2 - b 3 - a 4 - c 5 - b 6 - a 7 - b 8 - d 9 − c

Solution 2.2.4

1. $\sin x = \cos(-60°) = \cos 60° = \cos(90 - 30°)$

   => $\sin x = \sin 30°$ => $x = 30°$

2. $\sin 610° = \sin(360 + 250)° = \sin 250°$

   $= \sin(270 - 20)°$

   $= -\cos 20°$

3. $\cos 150 = \cos(180 - 30)°$

   $= -\cos 30° = -\dfrac{\sqrt{3}}{2}$

4. $\tan 1830 = \tan(5 \times 360 + 30)°$

   $= \tan 30° = \dfrac{1}{\sqrt{3}}$

5. $sin^2 52 + sin^2 38 = sin^2 52 + sin^2(90-52)$

$\qquad = sin^2 52 + cos^2 52 = 1$

6. $\sin 36 \ \sec 54 - \cos 24 \ \text{cosec } 66$

$\qquad = \sin 36 \ \sec(90-36) - \cos 24 \ \text{cosec}(90-24)$

$\qquad = \sin 36 \ \text{cosec } 36 - \cos 24 \ \sec 24$

$\qquad = \sin 36 \ \dfrac{1}{\sin 36} - \cos 24 \ \dfrac{1}{\cos 24} = 0$

7. $\dfrac{\sin(90-A)}{cosec(90-A)} + \dfrac{\cos(90-A)}{sec(90-A)} = \dfrac{\cos A}{\sec A} + \dfrac{\sin A}{\text{cosec } A}$

$\qquad\qquad\qquad\qquad = cos^2 A + sin^2 A = 1$

8. $1 - \dfrac{\sin(90-A)\sin A}{\tan A} = 1 - \cos A \sin A . \dfrac{\cos A}{\sin A}$

$\qquad\qquad\qquad = 1 - \cos^2 A$

$\qquad\qquad\qquad = sin^2 A$

9. $\cot(-675) = -\cot 675$

$\qquad\qquad = -\cot(360 + 315)$

$\qquad\qquad = -\cot 315$

$\qquad\qquad = \cot(360 - 45)$

$\qquad\qquad = \cot 45 = 1$

UNIT 3

3.1 Addition formulae

$\sin(A+B)$	$\sin(A-B)$
$= \sin A \cos B + \cos A \sin B$	$= \sin A \cos B - \cos A \sin B$
$\cos(A+B)$	$\cos(A-B)$
$= \cos A \cos B - \sin A \sin B$	$= \cos A \cos B + \sin A \sin B$
$\tan(A+B) = \dfrac{\tan A + \tan B}{1 - \tan A \tan B}$	$\tan(A-B) = \dfrac{\tan A - \tan B}{1 + \tan A \tan B}$
$\sin(A+B)\sin(A-B) = \sin^2 A - \sin^2 B$	
$\cos(A+B)\cos(A-B) = \cos^2 A - \sin^2 B$	

Examples 3.1

1. $\sin 75° = \sin(45° + 30°) = \sin 45° \cos 30° + \cos 45° \sin 30°$

$$= \frac{1}{\sqrt{2}} \cdot \frac{\sqrt{3}}{2} + \frac{1}{\sqrt{2}} \cdot \frac{1}{2}$$

$$= \frac{\sqrt{3}+1}{2\sqrt{2}}$$

2. $\cos 15° = \cos(45° - 30°)$

$= \cos 45° \cos 30° + \sin 45° \sin 30°$

$$= \frac{1}{\sqrt{2}} \cdot \frac{\sqrt{3}}{2} + \frac{1}{\sqrt{2}} \cdot \frac{1}{2}$$

$$= \frac{\sqrt{3}+1}{2\sqrt{2}}$$

Note : The values of $\sin 75°$ and $\cos 15°$ are equal. How?

$\sin 75° = \sin(90° - 15°) = \cos 15°$

3. $\cos A \, \cos(B - A) - \sin A \, \sin(B - A)$

$$= \cos(A + B - A) = \cos B$$

4. $\dfrac{\tan 69° + \tan 66°}{1 - \tan 69° . \tan 66°} = \tan(69° + 66°)$

$$= \tan(135°)$$

$$= \tan(90° + 45°)$$

$$= -\cot 45° = -1$$

5. $\dfrac{\cos(A-B) - \cos(A+B)}{\cos(A+B) + \cos(A-B)}$

$$= \dfrac{(\cos A \cos B + \sin A \sin B) - (\cos A \cos B - \sin A \sin B)}{\cos A \cos B - \sin A \sin B + \cos A \cos B + \sin A \sin B}$$

$$= \dfrac{2 \sin A \sin B}{2 \cos A \cos B}$$

$$= \tan A \, \tan B$$

Exercise 3.1 Choose the correct answer

1. $\sin 30 \, \cos 60 + \sin 60 \, \cos 30 = \ldots\ldots\ldots$

a) 0   b) $\dfrac{\sqrt{3}}{2}$ c) 1   d) $-\dfrac{1}{2}$

2. $\dfrac{\sin(A-B)}{\sin A \, \sin B} = \ldots\ldots\ldots$

a) $\cot A - \cot B$ b) $\tan A - \tan B$ c) $\cot B - \cot A$ d) $\tan A + \tan B$

3. $\cos(45 - A)\cos(45 - B) - \sin(45 - A)\sin(45 - B) = \ldots\ldots\ldots$

a) $\cos(A + B)$ b) $\sin(A + B)$ c) $\sin(A - B)$ d) $\cos(A - B)$

4. $\tan(\theta + 45) = \ldots\ldots\ldots$

a) $\dfrac{\tan \theta - 1}{1 + \tan \theta}$ b) $\dfrac{\tan \theta + 1}{1 - \tan \theta}$ c) $\dfrac{1 - \tan \theta}{1 + \tan \theta}$ d) $\dfrac{\tan \theta + 1}{\tan \theta - 1}$

5. $(\cos^2 B - \cos^2 A) - \sin(A+B)\sin(A-B) = $ ....

a) 1   b) 0   c) 2   d) −1

Answers :   1 – c   2 – c   3 – b   4 – b   5 – b

Solution 3.1

1. $\sin 30 \cos 60 + \sin 60 \cos 30$

$$= \sin(30 + 60) = \sin 90 = 1$$

2. $\dfrac{\sin(A-B)}{\sin A \sin B} = \dfrac{\sin A \cos B - \cos A \sin B}{\sin A \sin B}$

$$= \dfrac{\cos B}{\sin B} - \dfrac{\cos A}{\sin A}$$

$$= \cot B - \cot A$$

3. $\cos(45 - A)\cos(45 - B) - \sin(45 - A)\sin(45 - B)$

$$= \cos(45 - A + 45 - B)$$

$$= \cos[90 - (A + B)]$$

$$= \sin(A + B)$$

4. $\tan(\theta + 45) = \dfrac{\tan\theta + \tan 45}{1 - \tan 45 \tan\theta}$

$$= \dfrac{\tan\theta + 1}{1 - \tan\theta}$$

5. $(\cos^2 B - \cos^2 A) - \sin(A+B)\sin(A-B)$

$$= \cos^2 B - \cos^2 A - (\sin^2 A - \sin^2 B)$$

$$= (\cos^2 B + \sin^2 B) - (\cos^2 A + \sin^2 A)$$

$$= 1 - 1 = 0$$

Table 3.1

 **Try to recollect the following :**

sin θ cosec θ		
cos θ sec θ		
tan θ cot θ		
$\dfrac{1}{\sin\theta} = \csc\theta$ $\dfrac{1}{\cos\theta} = \sec\theta$ $\dfrac{1}{\tan\theta} = \cot\theta$	$\dfrac{1}{\csc\theta} = \sin\theta$ $\dfrac{1}{\sec\theta} = \cos\theta$ $\dfrac{1}{\cot\theta} = \tan\theta$	$\tan\theta = \dfrac{\sin\theta}{\cos\theta}$ $\cot\theta = \dfrac{\cos\theta}{\sin\theta}$
$\sin^2\theta + \cos^2\theta = 1$ $\sin^2\theta = 1 - \cos^2\theta$; $\cos^2\theta = 1 - \sin^2\theta$		
$1 + \tan^2\theta = \sec^2\theta$ $1 + \cot^2\theta = \csc^2\theta$	$\sec^2\theta - 1 = \tan^2\theta$ $\csc^2\theta - 1 = \cot^2\theta$	
$\sin(A+B)$ $= \sin A \cos B + \cos A \sin B$ $\cos(A+B)$ $= \cos A \cos B - \sin A \sin B$ $\tan(A+B) = \dfrac{\tan A + \tan B}{1 - \tan A \tan B}$	$\sin(A-B)$ $= \sin A \cos B - \cos A \sin B$ $\cos(A-B)$ $= \cos A \cos B + \sin A \sin B$ $\tan(A-B) = \dfrac{\tan A - \tan B}{1 + \tan A \tan B}$	
$\sin(A+B)\sin(A-B) = \sin^2 A - \sin^2 B$ $\cos(A+B)\cos(A-B) = \cos^2 A - \sin^2 B$		

3.2 Multiple angles

$$\sin 2A = \begin{cases} 2\sin A \cos A \\ \\ \dfrac{2\tan A}{1+\tan^2 A} \end{cases} \qquad \sin A = \begin{cases} 2\sin\dfrac{A}{2}\cos\dfrac{A}{2} \\ \\ \dfrac{2\tan\dfrac{A}{2}}{1+\tan^2\dfrac{A}{2}} \end{cases}$$

$$\cos 2A = \begin{cases} \cos^2 A - \sin^2 A \\ 2\cos^2 A - 1 \\ 1 - 2\sin^2 A \\ \dfrac{1-\tan^2 A}{1+\tan^2 A} \end{cases} \qquad \cos A = \begin{cases} \cos^2\dfrac{A}{2} - \sin^2\dfrac{A}{2} \\ 2\cos^2\dfrac{A}{2} - 1 \\ 1 - 2\sin^2\dfrac{A}{2} \\ \dfrac{1-\tan^2\dfrac{A}{2}}{1+\tan^2\dfrac{A}{2}} \end{cases}$$

$$\tan 2A = \dfrac{2\tan A}{1-\tan^2 A} \qquad \tan A = \dfrac{2\tan\dfrac{A}{2}}{1-\tan^2\dfrac{A}{2}}$$

$$1 + \cos 2A = 2\cos^2 A \qquad 1 + \cos A = 2\cos^2\dfrac{A}{2}$$

$$1 - \cos 2A = 2\sin^2 A \qquad 1 - \cos A = 2\sin^2\dfrac{A}{2}$$

$$\sin 3A = 3\sin A - 4\sin^3 A \qquad \sin 18° = \cos 72° = \dfrac{\sqrt{5}-1}{4}$$

$$\cos 3A = 4\cos^3 A - 3\cos A \qquad \cos 36° = \sin 54° = \dfrac{\sqrt{5}+1}{4}$$

$$\tan 3A = \dfrac{3\tan A - \tan^3 A}{1 - 3\tan^2 A}$$

Examples 3.2

1. $\sin 3A = \sin(2A + A)$

    $= \sin 2A \cos A + \cos 2A \sin A$

    $= 2\sin A \cos^2 A + (1 - 2\sin^2 A)\sin A$

    $= 2\sin A (1 - \sin^2 A) + \sin A - 2\sin^3 A$

    $= 3\sin A - 4\sin^3 A$

2. If $\tan A = 3$, find $\tan 3A$.

$$\tan 3A = \frac{3 \tan A - \tan^3 A}{1 - 3 \tan^2 A}$$

$$= \frac{9 - 27}{1 - 27} = \frac{9}{13}$$

3. $\tan A + \cot A = \dfrac{\sin A}{\cos A} + \dfrac{\cos A}{\sin A}$

$$= \frac{\sin^2 A + \cos^2 A}{\cos A \sin A} = \frac{1}{\cos A \sin A}$$

$$= \frac{2}{2 \sin A \cos A}$$

$$= \frac{2}{\sin 2A} = 2 \operatorname{cosec} 2A$$

4. If $\cos A = \dfrac{3}{5}$, find $\sin 2A$.

$$\sin A = \sqrt{1 - \cos^2 A} = \sqrt{1 - \frac{9}{25}} = \pm \frac{4}{5}$$

Hence, $\sin 2A = 2 \sin A \cos A$

$$= \pm 2 \cdot \frac{4}{5} \cdot \frac{3}{5} = \pm \frac{24}{25}$$

5. $\sqrt{1 - \sin A} = \sqrt{\sin^2 \frac{A}{2} + \cos^2 \frac{A}{2} - 2 \sin \frac{A}{2} \cos \frac{A}{2}}$

$$= \sqrt{(\sin \tfrac{A}{2} - \cos \tfrac{A}{2})^2}$$

$$= \sin \frac{A}{2} - \cos \frac{A}{2}$$

Exercise 3.2 : Choose the correct answer

1. $\dfrac{\sin 2A}{1 + \cos 2A} = \ldots\ldots\ldots\ldots$

a) $\sin A$ b) $\cos A$ c) $\tan A$ d) $\cot A$

2. $\dfrac{\sin 2A}{1-\cos 2A} = \ldots\ldots\ldots$

   a) sin A b) cos A c) tan A d) cot A

3. $\sqrt{\dfrac{1-\cos 2A}{1+\cos 2A}} = \ldots\ldots\ldots$

   a) sin A b) cos A c) tan A d) cot A

4. $\dfrac{\sin^3 A + \sin 3A}{3 \sin A} = \ldots\ldots\ldots$

   a) $\sin^2 A$ b) $\cos^2 A$ c) $\tan^2 A$ d) $\cot^2 A$

5. $\dfrac{2\cos^3 A + \cos 3A}{3 \cos A} = \ldots\ldots\ldots$

   a) sin 2A b) $-$cos 2A c) cos 2A d) $-$sin 2A

6. $1 + \cos 4A = \ldots\ldots\ldots$

   a) $2\sin^2 A$ b) $2\cos^2 A$ c) $2\cos^2 2A$ d) $2\sin^2 2A$

7. $\dfrac{3\cos A + \cos 3A}{3\sin A - \sin 3A} = \ldots\ldots\ldots$

   a) $\sin^3 A$ b) $\cos^3 A$ c) $\tan^3 A$ d) $\cot^3 A$

8. $\dfrac{\sin 3A}{\sin A} - \dfrac{\cos 3A}{\cos A} = \ldots\ldots\ldots$

   a) 0 b) 1 c) 2 d) 3

9. $\tan A + \cot A = \ldots\ldots\ldots$

   a) 2 sin 2A b) 2 cosec 2A c) 2 cos 2A d) 2 cot 2A

10. $\dfrac{1 - \tan^2(45-A)}{1 + \tan^2(45-A)} = \ldots\ldots\ldots$

    a) sin 2A b) $-$cos 2A c) cos 2A d) $-$sin 2A

11. $\dfrac{\sin A + \sin \frac{A}{2}}{1 + \cos A + \cos \frac{A}{2}} = \ldots\ldots\ldots$

a) $\sin \dfrac{A}{2}$ b) $\cos \dfrac{A}{2}$ c) $\tan \dfrac{A}{2}$ d) $\cot \dfrac{A}{2}$

12. $\dfrac{1-\cos A + \sin A}{1+ \cos A + \sin A} = \ldots\ldots\ldots$

a) $\sin \dfrac{A}{2}$ b) $\cos \dfrac{A}{2}$ c) $\tan \dfrac{A}{2}$ d) $\cot \dfrac{A}{2}$

Answers

1 - c   2 - d   3 - c   4 - b   5 - c   6 - c

7 - d   8 - c   9 - b   10 - a   11 - c   12 - c

Solution 3.2

1. $\dfrac{\sin 2A}{1+ \cos 2A} = \dfrac{2 \sin A \cos A}{2 \cos^2 A} = \tan A$

2. $\dfrac{\sin 2A}{1- \cos 2A} = \dfrac{2 \sin A \cos A}{2 \sin^2 A} = \cot A$

3. $\sqrt{\dfrac{1-\cos 2A}{1+\cos 2A}} = \sqrt{\dfrac{2 \sin^2 A}{2 \cos^2 A}} = \sqrt{\tan^2 A} = \tan A$

4. $\dfrac{\sin^3 A + \sin 3A}{3 \sin A} = \dfrac{\sin^3 A + 3 \sin A - 4 \sin^3 A}{3 \sin A}$

$= \dfrac{3 \sin A - 3 \sin^3 A}{3 \sin A} = \dfrac{3 \sin A (1 - \sin^2 A)}{3 \sin A}$

$= 1 - \sin^2 A = \cos^2 A$

5. $\dfrac{2 \cos^3 A + \cos 3A}{3 \cos A} = \dfrac{2 \cos^3 A + 4 \cos^3 A - 3 \cos A}{3 \cos A}$

$= \dfrac{3 \cos A (2 \cos^2 A - 1)}{3 \cos A}$

$= \cos 2A$

6. $1 + \cos 4A = 1 + \cos 2(2A)$

$= 1 + \cos 2\theta$ where $\theta = 2A$

$$= 2\cos^2\theta = 2\cos^2 2A$$

7. $\dfrac{3\cos A + \cos 3A}{3\sin A - \sin 3A} = \dfrac{4\cos^3 A}{4\sin^3 A} = \cot^3 A$

8. $\dfrac{\sin 3A}{\sin A} - \dfrac{\cos 3A}{\cos A} = \dfrac{3\sin A - 4\sin^3 A}{\sin A} - \dfrac{4\cos^3 A - 3\cos A}{\cos A}$

$$= 3 - 4\sin^2 A - 4\cos^2 A + 3$$

$$= 6 - 4(\sin^2 A + \cos^2 A)$$

$$= 6 - 4 = 2$$

9. $\tan A + \cot A = \dfrac{\sin A}{\cos A} + \dfrac{\cos A}{\sin A} = \dfrac{\sin^2 A + \cos^2 A}{\sin A \cos A}$

$$= \dfrac{1}{\sin A \cos A} = \dfrac{2}{2\sin A \cos A}$$

$$= \dfrac{2}{\sin 2A} = 2\operatorname{cosec} 2A$$

10. $\dfrac{1 - \tan^2(45 - A)}{1 + \tan^2(45 - A)} = \cos 2(45 - A) = \cos(90 - 2A) = \sin 2A$

11. $\dfrac{\sin A + \sin \frac{A}{2}}{1 + \cos A + \cos \frac{A}{2}} = \dfrac{2\sin \frac{A}{2}\cos \frac{A}{2} + \sin \frac{A}{2}}{2\cos^2 \frac{A}{2} + \cos \frac{A}{2}}$

$$= \dfrac{\sin \frac{A}{2}\left(2\cos \frac{A}{2} + 1\right)}{\cos \frac{A}{2}\left(2\cos \frac{A}{2} + 1\right)}$$

$$= \tan \dfrac{A}{2}$$

12. $\dfrac{1 - \cos A + \sin A}{1 + \cos A + \sin A} = \dfrac{2\sin^2 \frac{A}{2} + 2\sin \frac{A}{2}\cos \frac{A}{2}}{2\cos^2 \frac{A}{2} + 2\sin \frac{A}{2}\cos \frac{A}{2}}$

$$= \dfrac{2\sin \frac{A}{2}\left(\sin \frac{A}{2} + \cos \frac{A}{2}\right)}{2\cos \frac{A}{2}\left(\cos \frac{A}{2} + \sin \frac{A}{2}\right)} = \tan \dfrac{A}{2}$$

Table 3.2

   **Try to recollect the following**

$\sin\theta$ $\operatorname{cosec}\theta$ $\cos\theta$ $\sec\theta$ $\tan\theta$ $\cot\theta$	$\dfrac{1}{\sin\theta}=\operatorname{cosec}\theta$ $\dfrac{1}{\operatorname{cosec}\theta}=\sin\theta$ $\dfrac{1}{\cos\theta}=\sec\theta$ $\dfrac{1}{\sec\theta}=\cos\theta$ $\dfrac{1}{\tan\theta}=\cot\theta$ $\dfrac{1}{\cot\theta}=\tan\theta$	$\tan\theta=\dfrac{\sin\theta}{\cos\theta}$ $\cot\theta=\dfrac{\cos\theta}{\sin\theta}$
$\sin^2\theta+\cos^2\theta=1$; $\sin^2\theta=1-\cos^2\theta$; $\cos^2\theta=1-\sin^2\theta$		
$1+\tan^2\theta=\sec^2\theta$; $\sec^2\theta-1=\tan^2\theta$ $1+\cot^2\theta=\operatorname{cosec}^2\theta$; $\operatorname{cosec}^2\theta-1=\cot^2\theta$		

$\sin(A+B)=\sin A\cos B+\cos A\sin B$	$\sin(A-B)=\sin A\cos B-\cos A\sin B$
$\cos(A+B)=\cos A\cos B-\sin A\sin B$	$\cos(A-B)=\cos A\cos B+\sin A\sin B$
$\tan(A+B)=\dfrac{\tan A+\tan B}{1-\tan A\tan B}$	$\tan(A-B)=\dfrac{\tan A-\tan B}{1+\tan A\tan B}$
$\sin(A+B)\sin(A-B)=\sin^2 A-\sin^2 B$; $\cos(A+B)\cos(A-B)=\cos^2 A-\sin^2 B$	

$$\sin 2A = \begin{cases} 2\sin A\cos A \\[4pt] \dfrac{2\tan A}{1+\tan^2 A} \end{cases} \qquad \sin A = \begin{cases} 2\sin\dfrac{A}{2}\cos\dfrac{A}{2} \\[4pt] \dfrac{2\tan\dfrac{A}{2}}{1+\tan^2\dfrac{A}{2}} \end{cases}$$

$$\cos 2A = \begin{cases} \cos^2 A-\sin^2 A \\ 2\cos^2 A-1 \\ 1-2\sin^2 A \\ \dfrac{1-\tan^2 A}{1+\tan^2 A} \end{cases} \qquad \cos A = \begin{cases} \cos^2\dfrac{A}{2}-\sin^2\dfrac{A}{2} \\ 2\cos^2\dfrac{A}{2}-1 \\ 1-2\sin^2\dfrac{A}{2} \\ \dfrac{1-\tan^2\dfrac{A}{2}}{1+\tan^2\dfrac{A}{2}} \end{cases}$$

$$\tan 2A = \dfrac{2\tan A}{1-\tan^2 A} \qquad \tan A = \dfrac{2\tan\dfrac{A}{2}}{1-\tan^2\dfrac{A}{2}}$$

$$1+\cos 2A = 2\cos^2 A \qquad 1+\cos A = 2\cos^2\dfrac{A}{2}$$

$$1-\cos 2A = 2\sin^2 A \qquad 1-\cos A = 2\sin^2\dfrac{A}{2}$$

$$\sin 3A = 3\sin A-4\sin^3 A \qquad \sin 18°=\cos 72°=\dfrac{\sqrt{5}-1}{4}$$

$$\cos 3A = 4\cos^3 A-3\cos A \qquad \cos 36°=\sin 54°=\dfrac{\sqrt{5}+1}{4}$$

$$\tan 3A = \dfrac{3\tan A-\tan^3 A}{1-3\tan^2 A}$$

UNIT 4

4.1 Product formulae

$\sin C + \sin D = 2 \sin\frac{C+D}{2}\cos\frac{C-D}{2}$	$2 \sin A \cos B = \sin(A+B) + \sin(A-B)$
$\sin C - \sin D = 2 \cos\frac{C+D}{2}\sin\frac{C-D}{2}$	$2 \cos A \sin B = \sin(A+B) - \sin(A-B)$
$\cos C + \cos D = 2 \cos\frac{C+D}{2}\cos\frac{C-D}{2}$	$2 \cos A \cos B = \cos(A+B) + \cos(A-B)$
$\cos C - \cos D = -2 \sin\frac{C+D}{2}\sin\frac{C-D}{2}$	$2 \sin A \sin B = \cos(A-B) - \cos(A+B)$

Examples 4.1

1. $\sin 7A - \sin 5A = 2 \cos\frac{7A+5A}{2} \cdot \sin\frac{7A-5A}{2}$

$$= 2 \cos 6A \cdot \sin A$$

2. $\cos A + \cos 3A = 2 \cos\frac{A+3A}{2} \cdot \cos\frac{A-3A}{2}$

$$= 2 \cos 2A \cdot \cos(-A)$$

$$= 2 \cos 2A \cdot \cos A$$

3. $\dfrac{\cos B - \cos A}{\cos B + \cos A} = -\dfrac{2 \sin\frac{B+A}{2} \cdot \sin\frac{B-A}{2}}{2 \cos\frac{B+A}{2} \cdot \cos\frac{B-A}{2}}$

$$= -\tan\frac{B+A}{2} \cdot \tan\frac{B-A}{2}$$

4. $\cos 2A \cdot \cos A = \frac{1}{2}(2 \cos 2A \cdot \cos A)$

$$= \frac{1}{2}(\cos 3A + \cos A)$$

5. $2 \sin 5A \cdot \sin 3A = \cos 2A - \cos 8A$

Exercise 4.1 : Choose the correct answer

1. sin 6A + sin 4A = 2......................

   a) sin 5A cos 2A   b) sin 5A cos A   c) sin A cos 2A   d) cos 5A sin A

2. cos 3A − cos 7A = 2......................

   a) sin 5A sin 2A   b) cos 5A sin 2A   c) sin 5A cos 2A   d) cos 5A cos 2A

3. $\dfrac{\sin A + \sin B}{\cos A + \cos B}$ = ...

   a) cos $(\dfrac{A+B}{2})$ b) sin $(\dfrac{A+B}{2})$ c) tan $(\dfrac{A+B}{2})$ d) cot $(\dfrac{A+B}{2})$

4. $\dfrac{\sin 5A - \sin 3A}{\cos 3A + \cos 5A}$ = ...

   a) sin A   b) cos A   c) tan A   d) cot A

5. $\dfrac{\cos 21° - \sin 21°}{\cos 21° + \sin 21°}$ = ...

   a) sin 24°   b) cos 24°   c) cot 24°   d) tan 24°

6. $\dfrac{\sin 75° - \sin 15°}{\cos 75° + \cos 15°}$ = ...

   a) $\dfrac{1}{\sqrt{2}}$ b) $\dfrac{1}{\sqrt{3}}$ c) $\dfrac{1}{2}$ d) $\dfrac{1}{3}$

7. 2 sin 70° sin 20° = ...

   a) sin 50°   b) cos 50°   c) 1   d) 0

8. 2 sin (45 + A) sin (45 − A) = ...

   a) cos 2A   b) tan 2A   c) sin 2A   d) cot 2A

9. cos 70° + cos 50° = ...

   a) cos 60°   b) sin 60°   c) sin 10°   d) cos 10°

10. If A + B + C = π, then cos (A − B) + cos C = ...

a) 2 sin A sin B   b) 2 sin A cos B   c) sin A sin B   d) sin A cos B

Answers: 1 - b   2 - a   3 - c   4 - c   5 - d

   6 - b   7 - b   8 - a   9 - d   10 - a

Solution 4.1

1. $\sin 6A + \sin 4A = 2 \sin \dfrac{10A}{2} \cdot \cos \dfrac{2A}{2}$

   $= 2 \sin 5A \cdot \cos A$

2. $\cos 3A - \cos 7A = -2 \sin \dfrac{10A}{2} \cdot \sin(-\dfrac{4A}{2})$

   $= 2 \sin 5A \cdot \sin 2A$

3. $\dfrac{\sin A + \sin B}{\cos A + \cos B} = \dfrac{2 \sin \dfrac{A+B}{2} \cos \dfrac{A-B}{2}}{2 \cos \dfrac{A+B}{2} \cos \dfrac{A-B}{2}}$

   $= \tan \dfrac{A+B}{2}$

4. $\dfrac{\sin 5A - \sin 3A}{\cos 3A + \cos 5A} = \dfrac{2 \cos \dfrac{8A}{2} \sin \dfrac{2A}{2}}{2 \cos \dfrac{8A}{2} \cos \dfrac{2A}{2}}$

   $= \tan A$

5. $\dfrac{\cos 21^o - \sin 21^o}{\cos 21^o + \sin 21^o} = \dfrac{1 - \tan 21^o}{1 + \tan 21^o}$ (dividing by $\cos 21^o$)

   $= \dfrac{\tan 45^o - \tan 21^o}{1 + \tan 45^o \tan 21^o}$

   $= \tan(45^o - 21^o) = \tan 24^o$

6. $\dfrac{\sin 75^o - \sin 15^o}{\cos 75^o + \cos 15^o} = \dfrac{2 \cos 45^o \sin 30^o}{2 \cos 45^o \cos 30^o}$

   $= \tan 30^o = \dfrac{1}{\sqrt{3}}$

7. $2 \sin 70^o \cdot \sin 20^o = \cos(70^o - 20^o) - \cos(70^o + 20^o)$

$$= \cos 50° - \cos 90° = \cos 50°$$

8. $2 \sin(45 + A) \cdot \sin(45 - A)$

$$= \cos[(45 + A) - (45 - A)] - \cos[(45 + A) + (45 - A)]$$

$$= \cos 2A - \cos 90° = \cos 2A$$

9. $\cos 70° + \cos 50°$

$$= 2 \cos \frac{70+50}{2} \cos \frac{70-50}{2}$$

$$= 2 \cos 60 \cos 10$$

$$= 2 \cdot \frac{1}{2} \cdot \cos 10 = \cos 10$$

10. $A + B + C = \pi \Rightarrow C = \pi - (A + B)$

$\cos(A - B) + \cos C = \cos(A - B) + \cos[\pi - (A + B)]$

$$= \cos(A - B) - \cos(A + B)$$

$$= \cos A \cos B + \sin A \sin B - \cos A \cos B + \sin A \sin B$$

$$= 2 \sin A \sin B$$

 **Try to recollect the following : Table I**

$\sin\theta$ $\csc\theta$ $\cos\theta$ $\sec\theta$ $\tan\theta$ $\cot\theta$	$\frac{1}{\sin\theta} = \csc\theta$ $\frac{1}{\sec\theta} = \cos\theta$	$\frac{1}{\csc\theta} = \sin\theta$ $\frac{1}{\tan\theta} = \cot\theta$	$\frac{1}{\cos\theta} = \sec\theta$ $\frac{1}{\cot\theta} = \tan\theta$	$\tan\theta = \frac{\sin\theta}{\cos\theta}$ $\cot\theta = \frac{\cos\theta}{\sin\theta}$
$\sin^2\theta + \cos^2\theta = 1$;		$\sin^2\theta = 1 - \cos^2\theta$;	$\cos^2\theta = 1 - \sin^2\theta$	
	$1 + \tan^2\theta = \sec^2\theta$	;	$\sec^2\theta - 1 = \tan^2\theta$	
	$1 + \cot^2\theta = \csc^2\theta$	;	$\csc^2\theta - 1 = \cot^2\theta$	

$\sin(A + B) = \sin A \cos B + \cos A \sin B$	$\sin(A - B) = \sin A \cos B - \cos A \sin B$
$\cos(A + B) = \cos A \cos B - \sin A \sin B$	$\cos(A - B) = \cos A \cos B + \sin A \sin B$
$\tan(A + B) = \dfrac{\tan A + \tan B}{1 - \tan A \tan B}$	$\tan(A - B) = \dfrac{\tan A - \tan B}{1 + \tan A \tan B}$

$$\sin(A+B)\sin(A-B) = \sin^2 A - \sin^2 B$$
$$\cos(A+B)\cos(A-B) = \cos^2 A - \sin^2 B$$

$$\sin 2A = \begin{cases} 2\sin A \cos A \\ \dfrac{2\tan A}{1+\tan^2 A} \end{cases} \qquad \sin A = \begin{cases} 2\sin\dfrac{A}{2}\cos\dfrac{A}{2} \\ \dfrac{2\tan\dfrac{A}{2}}{1+\tan^2\dfrac{A}{2}} \end{cases}$$

$$\cos 2A = \begin{cases} \cos^2 A - \sin^2 A \\ 2\cos^2 A - 1 \\ 1 - 2\sin^2 A \\ \dfrac{1-\tan^2 A}{1+\tan^2 A} \end{cases} \qquad \cos A = \begin{cases} \cos^2\dfrac{A}{2} - \sin^2\dfrac{A}{2} \\ 2\cos^2\dfrac{A}{2} - 1 \\ 1 - 2\sin^2\dfrac{A}{2} \\ \dfrac{1-\tan^2\dfrac{A}{2}}{1+\tan^2\dfrac{A}{2}} \end{cases}$$

$$\tan 2A = \dfrac{2\tan A}{1-\tan^2 A} \qquad \tan A = \dfrac{2\tan\dfrac{A}{2}}{1-\tan^2\dfrac{A}{2}}$$

$$1 + \cos 2A = 2\cos^2 A \qquad 1 + \cos A = 2\cos^2\dfrac{A}{2}$$

$$1 - \cos 2A = 2\sin^2 A \qquad 1 - \cos A = 2\sin^2\dfrac{A}{2}$$

$$\sin 3A = 3\sin A - 4\sin^3 A \qquad \sin 18° = \cos 72° = \dfrac{\sqrt{5}-1}{4}$$

$$\cos 3A = 4\cos^3 A - 3\cos A \qquad \cos 36° = \sin 54° = \dfrac{\sqrt{5}+1}{4}$$

$$\tan 3A = \dfrac{3\tan A - \tan^3 A}{1 - 3\tan^2 A}$$

$$\sin C + \sin D = 2\sin\dfrac{C+D}{2}\cos\dfrac{C-D}{2}$$
$$\sin C - \sin D = 2\cos\dfrac{C+D}{2}\sin\dfrac{C-D}{2}$$
$$\cos C + \cos D = 2\cos\dfrac{C+D}{2}\cos\dfrac{C-D}{2}$$
$$\cos C - \cos D = -2\sin\dfrac{C+D}{2}\cos\dfrac{C-D}{2}$$

$$2\sin A \cos B = \sin(A+B) + \sin(A-B)$$
$$2\cos A \sin B = \sin(A+B) - \sin(A-B)$$
$$2\cos A \cos B = \cos(A+B) + \cos(A-B)$$
$$2\sin A \sin B = \cos(A-B) - \cos(A+B)$$

UNIT 5

5.1 Trigonometric equations

The principal values of sine function lies in $[-\frac{\pi}{2}, \frac{\pi}{2}]$, cosine function lies in $[0, \pi]$ and tan lies in $(-\frac{\pi}{2}, \frac{\pi}{2})$	
$\sin\theta = 0 \Rightarrow \theta = n\pi, n \in Z$	$\sin\theta = \sin\alpha \Rightarrow \theta = n\pi + (-1)^n \alpha,\ n \in Z$
$\cos\theta = 0 \Rightarrow \theta = (2n+1)\frac{\pi}{2}, n \in Z$	$\cos\theta = \cos\alpha \Rightarrow \theta = 2n\pi \pm \alpha,\quad n \in Z$
$\tan\theta = 0 \Rightarrow \theta = n\pi, n \in Z$	$\tan\theta = \tan\alpha \Rightarrow \theta = n\pi + \alpha,\quad n \in Z$
If $a\cos\theta + b\sin\theta = c$, $|c| \leq \sqrt{a^2 + b^2}$, then $\theta = \alpha + 2n\pi \pm \cos^{-1}\frac{c}{r}$ where $r = \sqrt{a^2+b^2}$ and $\alpha = \tan^{-1}\frac{b}{a}$	

Examples 5.1

1. $\sin x = \frac{\sqrt{3}}{2}$ $\Rightarrow \sin x = \sin\frac{\pi}{3}$ $\Rightarrow x = n\pi + (-1)^n \frac{\pi}{3}$

2. $\cos x = -\frac{1}{2}$ $\Rightarrow \cos x = \cos\frac{2\pi}{3}$ $\Rightarrow x = 2n\pi \pm \frac{2\pi}{3}$

3. $\tan^2 x = \frac{1}{3}$ $\Rightarrow \tan x = \pm\frac{1}{\sqrt{3}}$

   $\tan x = \frac{1}{\sqrt{3}}$ $\Rightarrow \tan x = \tan\frac{\pi}{6}$ $\Rightarrow x = n\pi + \frac{\pi}{6}$ ..... (1)

   $\tan x = -\frac{1}{\sqrt{3}}$ $\Rightarrow \tan x = \tan(-\frac{\pi}{6})$ $\Rightarrow x = n\pi + (-\frac{\pi}{6})$ .... (2)

   (1) & (2) $\Rightarrow x = n\pi \pm \frac{\pi}{6}$

4. Solve $\tan 5x = \cot 3x$

   Solution $\tan 5x = \tan(\frac{\pi}{2} - 3x)$

   $\Rightarrow 5x = n\pi + \frac{\pi}{2} - 3x$

   $\Rightarrow x = \frac{1}{8}(n\pi + \frac{\pi}{2})$ where n is an integer.

5. $\sqrt{3} \cos x + \sin x = \sqrt{2}$ .... (1)

Here $a = \sqrt{3}; b = 1; c = \sqrt{2}$. Also $|c| \leq \sqrt{a^2 + b^2}$

Dividing (1) by $\sqrt{a^2 + b^2} = 2$, we get

$\dfrac{\sqrt{3}}{2} \cos x + \dfrac{1}{2} \sin x = \dfrac{1}{\sqrt{2}}$ => $\sin \dfrac{\pi}{3} \cos x + \cos \dfrac{\pi}{3} \sin x = \dfrac{1}{\sqrt{2}}$

=> $\sin\left(x + \dfrac{\pi}{3}\right) = \sin \dfrac{\pi}{4}$

=> $x + \dfrac{\pi}{3} = n\pi + (-1)^n \dfrac{\pi}{4}$

=> $x = n\pi + (-1)^n \dfrac{\pi}{4} - \dfrac{\pi}{3}$ n is an integer.

Exercise 5.1 : Choose the correct answer

1. If $\sin 4x = 0$, then $x =$ ....

a) $n\pi$ b) $2n\pi$ c) $\dfrac{n\pi}{2}$ d) $\dfrac{n\pi}{4}$

2. If $\tan \dfrac{x}{2} = 0$, then $x =$ ....

a) $n\pi$ b) $2n\pi$ c) $\dfrac{n\pi}{2}$ d) $\dfrac{n\pi}{4}$

3. If $\tan \dfrac{x}{2} = 1$, then $x =$ ....

a) $n\pi$ b) $2n\pi + \dfrac{\pi}{2}$ c) $\dfrac{n\pi}{2}$ d) $\dfrac{n\pi}{4}$

4. If $\cos 2x = \dfrac{1}{2}$, then $x =$ ....

a) $n\pi \pm \dfrac{\pi}{6}$ b) $2n\pi + \dfrac{\pi}{2}$ c) $\dfrac{n\pi}{2}$ d) $\dfrac{n\pi}{4}$

5. If $\sin x = \dfrac{1}{2}$, then $x =$ ....

a) $n\pi + (-1)^n \dfrac{\pi}{6}$ b) $2n\pi + (-1)^n \dfrac{\pi}{6}$ c) $\dfrac{n\pi}{2}$ d) $\dfrac{n\pi}{4}$

6. If $\cos x = 1$, then $x = $ ....

a) $n\pi$ b) $2n\pi$ c) $\dfrac{n\pi}{2}$ d) $\dfrac{n\pi}{4}$

7. If $\cos x = -1$, then $x = $ ....

a) $n\pi$ b) $2n\pi + \pi$ c) $2n\pi - \pi$ d) b & c

8. If $\sin^2 x = \dfrac{1}{4}$, then $x = $ ....

a) $n\pi + (-1)^n \dfrac{\pi}{2}$ b) $n\pi \pm \dfrac{\pi}{6}$ c) $2n\pi - \dfrac{\pi}{2}$ d) $\dfrac{n\pi}{4}$

9. If $\sin x + \cos x = \sqrt{2}$, then $x = $ ....

a) $2n\pi + \dfrac{\pi}{2}$ b) $2n\pi + \dfrac{\pi}{4}$ c) $2n\pi + \dfrac{\pi}{6}$ d) $2n\pi$

10. If $\cos x + \sqrt{3}\sin x = 2$, then $x = 2n\pi + $ ....

a) $\dfrac{\pi}{2}$ b) $\dfrac{\pi}{6}$ c) $\dfrac{\pi}{3}$ d) $\dfrac{\pi}{4}$

Answers : 1 – d 2 – b 3 – b 4 – a 5 – a
 6 – b 7 – d 8 – b 9 – b 10 – c

Solution 5.1

1. $\sin 4x = 0 \quad \Rightarrow \quad 4x = n\pi \quad \Rightarrow \quad x = \dfrac{n\pi}{4}$

2. $\tan \dfrac{x}{2} = 0 \quad \Rightarrow \quad \dfrac{x}{2} = n\pi \quad \Rightarrow \quad x = 2n\pi$

3. $\tan \dfrac{x}{2} = 1 \quad \Rightarrow \quad \tan \dfrac{x}{2} = \tan \dfrac{\pi}{4}$

$\quad \Rightarrow \quad \dfrac{x}{2} = n\pi + \dfrac{\pi}{4} \quad \Rightarrow \quad x = 2n\pi + \dfrac{\pi}{2}$

4. $\cos 2x = \dfrac{1}{2} \quad \Rightarrow \quad \cos 2x = \cos \dfrac{\pi}{3}$

$$\Rightarrow 2x = 2n\pi \pm \frac{\pi}{3} \Rightarrow x = n\pi \pm \frac{\pi}{6}$$

5. $\sin x = \frac{1}{2}$ $\Rightarrow \sin x = \sin \frac{\pi}{6}$

$$\Rightarrow x = n\pi + (-1)^n \frac{\pi}{6}$$

6. $\cos x = 1 \Rightarrow \cos x = \cos 0$

$$\Rightarrow x = 2n\pi \pm 0 \Rightarrow x = 2n\pi$$

7. $\cos x = -1$ $\Rightarrow \cos x = \cos \pi$

$$\Rightarrow x = 2n\pi \pm \pi$$

8. $\sin^2 x = \frac{1}{4}$ $\Rightarrow \sin x = \pm \frac{1}{2}$

$\sin x = \frac{1}{2}$ $\Rightarrow \sin x = \sin \frac{\pi}{6}$

$$\Rightarrow x = n\pi + (-1)^n \frac{\pi}{6} \quad \ldots\ldots\ldots\ldots (1)$$

$\sin x = -\frac{1}{2}$ $\Rightarrow \sin x = \sin(-\frac{\pi}{6})$

$$\Rightarrow x = n\pi + (-1)^n (-\frac{\pi}{6}) \quad \ldots\ldots\ldots\ldots (2)$$

(1) & (2) $\Rightarrow x = n\pi \pm \frac{\pi}{6}$

9. $\sin x + \cos x = \sqrt{2}$ ……………. (1)

Here $a = 1; b = 1; c = \sqrt{2}$. Also $|c| \leq \sqrt{a^2 + b^2}$

Dividing (1) by $\sqrt{a^2 + b^2} = \sqrt{2}$, we get

$\frac{1}{\sqrt{2}} \sin x + \frac{1}{\sqrt{2}} \cos x = 1$ $\Rightarrow \sin \frac{\pi}{4} \sin x + \cos \frac{\pi}{4} \cos x = 1$

$$\Rightarrow \cos(x - \frac{\pi}{4}) = \cos 0$$

$$\Rightarrow x - \frac{\pi}{4} = 2n\pi \pm 0$$

$$\Rightarrow x = 2n\pi + \frac{\pi}{4} \quad \text{n is any integer.}$$

10. $\cos x + \sqrt{3} \sin x = 2$

Here $a = 1; b = \sqrt{3}; c = 2$. Also $|c| \leq \sqrt{a^2 + b^2}$

Dividing (1) by $\sqrt{a^2 + b^2} = 2$, we get

$$\frac{1}{2} \cos x + \frac{\sqrt{3}}{2} \sin x = 1 \quad \Rightarrow \cos \frac{\pi}{3} \cos x + \sin \frac{\pi}{3} \sin x = 1$$

$$\Rightarrow \cos\left(x - \frac{\pi}{3}\right) = \cos 0$$

$$\Rightarrow x - \frac{\pi}{3} = 2n\pi \pm 0$$

$$\Rightarrow x = 2n\pi + \frac{\pi}{3} \quad \text{n is any integer.}$$

Table 5.1

☺ **Try to recollect the following :**

The principal value of sine function lies in $[-\frac{\pi}{2}, \frac{\pi}{2}]$, cosine function lies in $[0, \pi]$ and tan lies in $(-\frac{\pi}{2}, \frac{\pi}{2})$	
$\sin \theta = 0 \Rightarrow \theta = n\pi, n \in Z$	$\sin \theta = \sin \alpha \Rightarrow \theta = n\pi + (-1)^n \alpha, n \in Z$
$\cos \theta = 0 \Rightarrow \theta = (2n+1)\frac{\pi}{2}, n \in Z$	$\cos \theta = \cos \alpha \Rightarrow \theta = 2n\pi \pm \alpha, \quad n \in Z$
$\tan \theta = 0 \Rightarrow \theta = n\pi, n \in Z$	$\tan \theta = \tan \alpha \Rightarrow \theta = n\pi + \alpha, \quad n \in Z$
If $a \cos \theta + b \sin \theta = c$, $|c| \leq \sqrt{a^2 + b^2}$, then $\theta = \alpha + 2n\pi \pm \cos^{-1}\frac{c}{r}$ where $r = \sqrt{a^2 + b^2}$ and $\alpha = \tan^{-1}\frac{b}{a}$	

UNIT 6

6.1 Inverse trigonometric functions

For suitable values of domain, we have

$\sin^{-1}(\sin x) = x$	$\cos^{-1}(\cos x) = x$	$\tan^{-1}(\tan x) = x$
$\operatorname{cosec}^{-1}(\operatorname{cosec} x) = x$	$\sec^{-1}(\sec x) = x$	$\cot^{-1}(\cot x) = x$
$\sin^{-1}\frac{1}{x} = \operatorname{cosec}^{-1} x$	$\cos^{-1}\frac{1}{x} = \sec^{-1} x$	$\tan^{-1}\frac{1}{x} = \cot^{-1} x$
$\operatorname{cosec}^{-1}\frac{1}{x} = \sin^{-1} x$	$\sec^{-1}\frac{1}{x} = \cos^{-1} x$	$\cot^{-1}\frac{1}{x} = \tan^{-1} x$
$\sin^{-1}(-x) = -\sin^{-1} x$ and $\cos^{-1}(-x) = \pi - \cos^{-1} x$ for $-1 \le x \le 1$		
$\sec^{-1}(-x) = \pi - \sec^{-1} x$ and $\tan^{-1}(-x) = -\tan^{-1} x$ for all real x.		
$\cot^{-1}(-x) = -\cot^{-1} x$, $\operatorname{cosec}^{-1}(-x) = -\operatorname{cosec}^{-1} x$		

$\sin^{-1} x + \cos^{-1} x = \frac{\pi}{2}$, $-1 \le x \le 1$	$\tan^{-1} x + \tan^{-1} y = \tan^{-1}\frac{x+y}{1-xy}$, $xy < 1$
$\operatorname{cosec}^{-1} x + \sec^{-1} x = \frac{\pi}{2}$	$\tan^{-1} x - \tan^{-1} y = \tan^{-1}\frac{x-y}{1+xy}$, $xy < 1$

$\sin^{-1} x = \cos^{-1}\sqrt{1-x^2}$, $0 \le x \le 1$; $\sin^{-1} x = -\cos^{-1}\sqrt{1-x^2}$, $-1 \le x \le 0$

If a, b and the radicals are positive, then $\sin^{-1}\frac{a}{\sqrt{a^2+b^2}} = \cos^{-1}\frac{b}{\sqrt{a^2+b^2}} = \tan^{-1}\frac{a}{b}$

Examples 6.1

1. Find the principal value of $\sin^{-1}\frac{1}{2}$

Let $\sin^{-1}\frac{1}{2} = y \Rightarrow \sin y = \frac{1}{2}$

The principal value of $\sin^{-1}$ is $[-\frac{\pi}{2}, \frac{\pi}{2}]$ and $\sin\frac{\pi}{6} = \frac{1}{2}$,

$\Rightarrow y = \frac{\pi}{6}$

2. Find the principal value of $\operatorname{cosec}^{-1}(-2)$

Let $\operatorname{cosec}^{-1}(-2) = y \Rightarrow \operatorname{cosec} y = -2$

The principal value of $cosec^{-1}$ is $[-\frac{\pi}{2}, \frac{\pi}{2}] - \{0\}$ and

$cosec\left(-\frac{\pi}{6}\right) = -2 \Rightarrow y = -\frac{\pi}{6}$

3. $tan^{-1}x + cot^{-1}x = \frac{\pi}{2}$

Let $\theta = tan^{-1}x \Rightarrow x = tan\,\theta = cot\left(\frac{\pi}{2} - \theta\right)$

$\Rightarrow cot^{-1}x = \frac{\pi}{2} - \theta$

$\Rightarrow \theta + cot^{-1}x = \frac{\pi}{2}$

$\Rightarrow tan^{-1}x + cot^{-1}x = \frac{\pi}{2}$

4. $2\,tan^{-1}x = sin^{-1}\frac{2x}{1+x^2}, \quad -1 \le x \le 1$

Let $\theta = tan^{-1}x \Rightarrow x = tan\,\theta$

R.H.S $= sin^{-1}\frac{2x}{1+x^2} = sin^{-1}\frac{2\,tan\,\theta}{1+tan^2\theta}$

$= sin^{-1}(sin\,2\theta) = 2\theta = 2\,tan^{-1}x$

Similarly, $2\,tan^{-1}x = cos^{-1}\frac{1-x^2}{1+x^2}, \quad x \ge 0$

and $2\,tan^{-1}x = tan^{-1}\frac{2x}{1-x^2}, \quad -1 < x < 1$

5. Find the value of $sin^{-1}\left(sin\,\frac{3\pi}{5}\right)$

Clearly $sin^{-1}\left(sin\,\frac{3\pi}{5}\right) = \frac{3\pi}{5}$.

But $\frac{3\pi}{5}$ lies outside the principal range of $sin^{-1}x$.

$\sin \frac{3\pi}{5} = \sin(\pi - \frac{2\pi}{5}) = \sin \frac{2\pi}{5}$ and $\frac{2\pi}{5}$ lies in the principal range of $\sin^{-1} x$.

Hence $\sin^{-1}(\sin \frac{3\pi}{5}) = \frac{2\pi}{5}$

Exercise 6.1 : Choose the correct answer

1. $\cos^{-1} \frac{1}{2} = $ ....

a) $\frac{\pi}{2}$ b) $\frac{\pi}{3}$ c) $\frac{2\pi}{3}$ d) $\frac{\pi}{4}$

2. $\tan^{-1}(-1) = $ ....

a) $-\frac{\pi}{2}$ b) $-\frac{\pi}{3}$ c) $-\frac{2\pi}{3}$ d) $-\frac{\pi}{4}$

3. $\sec^{-1} \frac{2}{\sqrt{3}} = $ ....

a) $\frac{\pi}{6}$ b) $\frac{\pi}{3}$ c) $\frac{2\pi}{3}$ d) $\frac{\pi}{4}$

4. $\tan^{-1} x + \tan^{-1} \frac{1}{x} = $ ....

a) $\frac{\pi}{2}$ b) $\frac{\pi}{3}$ c) $\frac{2\pi}{3}$ d) $\frac{\pi}{4}$

5. $\cos^{-1} \frac{2}{3} + \sin^{-1} \frac{2}{3} = $ ....

a) $\frac{\pi}{6}$ b) $\frac{\pi}{2}$ c) $\frac{2\pi}{3}$ d) $\frac{\pi}{4}$

6. $\cos^{-1} \frac{5}{13} = $ ....

a) $\cos^{-1} \frac{12}{13}$ b) $\sin^{-1} \frac{5}{13}$ c) $\sin^{-1} \frac{12}{13}$ d) $\frac{12}{13}$

7. $\tan^{-1}(\tan \frac{\pi}{3}) = $ ....

a) $\frac{\pi}{6}$ b) $\frac{\pi}{2}$ c) $\frac{\pi}{3}$ d) $\frac{\pi}{4}$

8. $\sin(\sin^{-1}x) = $ .... $x \in [-1, 1]$

   a) x b) $-x$ c) $\pi - x$ d) $\pi + x$

9. $\tan^{-1}[\tan(\cos^{-1}\frac{\sqrt{3}}{2})] = $ ....

   a) $\frac{\pi}{6}$ b) $\frac{\pi}{2}$ c) $\frac{\pi}{3}$ d) $\frac{\pi}{4}$

10. $2\tan^{-1}(\frac{1}{3}) = $ ....

    a) $\tan^{-1}\frac{1}{3}$ b) $\tan^{-1}\frac{2}{3}$ c) $\tan^{-1}\frac{3}{4}$ d) $\tan^{-1}\frac{1}{4}$

Answers :   1 - b   2 - d   3 - a   4 - a   5 - b

   6 - c   7 - c   8 - a   9 - a   10 - c

Solution 6.1

1. Let $\cos^{-1}\frac{1}{2} = y \Rightarrow \cos y = \frac{1}{2}$

   As the principal value of $\cos^{-1}$ is $[0, \pi]$ and $\cos\frac{\pi}{3} = \frac{1}{2}$,

   $$\Rightarrow y = \frac{\pi}{3}$$

2. Let $\tan^{-1}(-1) = y \Rightarrow \tan y = -1$

   As the principal value of $\tan^{-1}$ is $(-\frac{\pi}{2}, \frac{\pi}{2})$ and $\tan(-\frac{\pi}{4}) = -1$,

   $$\Rightarrow y = -\frac{\pi}{4}$$

3. Let $\sec^{-1}\frac{2}{\sqrt{3}} = y \Rightarrow \sec y = \frac{2}{\sqrt{3}}$

   As the principal value of $\sec^{-1}$ is $[0, \pi] - \{\frac{\pi}{2}\}$ and $\sec\frac{\pi}{6} = \frac{2}{\sqrt{3}}$,

   $$\Rightarrow y = \frac{\pi}{6}$$

4. $\tan^{-1}x + \tan^{-1}\frac{1}{x} = \tan^{-1}x + \cot^{-1}x = \frac{\pi}{2}$

5. $\cos^{-1}\frac{2}{3} + \sin^{-1}\frac{2}{3} = \frac{\pi}{2}$ as $-1 \leq \frac{2}{3} \leq 1$

6. $\cos^{-1}\frac{5}{13} = x \quad \Rightarrow \cos x = \frac{5}{13}$

$$\Rightarrow \sqrt{1 - \sin^2 x} = \frac{5}{13}$$

$$\Rightarrow \sin^2 x = 1 - \frac{25}{169} = \frac{144}{169}$$

$$\Rightarrow \sin x = \frac{12}{13}$$

$$\Rightarrow x = \sin^{-1}\left(\frac{12}{13}\right)$$

7. $\tan^{-1}(\tan\frac{\pi}{3}) = \frac{\pi}{3}$

8. $\sin(\sin^{-1}x) = x$

9. $\tan^{-1}[\tan(\cos^{-1}\frac{\sqrt{3}}{2})] = \tan^{-1}[\tan(\cos^{-1}(\cos\frac{\pi}{6}))]$

$$= \tan^{-1}[\tan(\frac{\pi}{6})] = \frac{\pi}{6}$$

10. $2\tan^{-1}(\frac{1}{3}) = \tan^{-1}(\frac{1}{3}) + \tan^{-1}(\frac{1}{3})$

$$= \tan^{-1}\left(\frac{\frac{1}{3}+\frac{1}{3}}{1-\frac{1}{3}\cdot\frac{1}{3}}\right)$$

$$= \tan^{-1}\left(\frac{\frac{2}{3}}{1-\frac{1}{9}}\right)$$

$$= \tan^{-1}\left(\frac{3}{4}\right)$$

Table 6.1  Try to recollect the following Table:

The principal value of sine function lies in $[-\frac{\pi}{2}, \frac{\pi}{2}]$, cosine function lies in $[0, \pi]$ and tan lies in $(-\frac{\pi}{2}, \frac{\pi}{2})$	
$\sin\theta = 0 \Rightarrow \theta = n\pi, n \in Z$	$\sin\theta = \sin\alpha \Rightarrow \theta = n\pi + (-1)^n \alpha, n \in Z$
$\cos\theta = 0 \Rightarrow \theta = (2n+1)\frac{\pi}{2}, n \in Z$	$\cos\theta = \cos\alpha \Rightarrow \theta = 2n\pi \pm \alpha, \quad n \in Z$
$\tan\theta = 0 \Rightarrow \theta = n\pi, n \in Z$	$\tan\theta = \tan\alpha \Rightarrow \theta = n\pi + \alpha, \quad n \in Z$
If $a\cos\theta + b\sin\theta = c$, $|c| \le \sqrt{a^2+b^2}$, then $\theta = \alpha + 2n\pi \pm \cos^{-1}\frac{c}{r}$ where $r = \sqrt{a^2+b^2}$ and $\alpha = \tan^{-1}\frac{b}{a}$	
$\sin^{-1}(\sin x) = x \qquad \cos^{-1}(\cos x) = x \qquad \tan^{-1}(\tan x) = x$ $\text{cosec}^{-1}(\text{cosec } x) = x \qquad \sec^{-1}(\sec x) = x \qquad \cot^{-1}(\cot x) = x$	
$\sin^{-1}\frac{1}{x} = \text{cosec}^{-1} x \qquad \cos^{-1}\frac{1}{x} = \sec^{-1} x \qquad \tan^{-1}\frac{1}{x} = \cot^{-1} x$ $\text{cosec}^{-1}\frac{1}{x} = \sin^{-1} x \qquad \sec^{-1}\frac{1}{x} = \cos^{-1} x \qquad \cot^{-1}\frac{1}{x} = \tan^{-1} x$	
$\sin^{-1}(-x) = -\sin^{-1}x \quad$ and $\quad \cos^{-1}(-x) = \pi - \cos^{-1}x \;$ for $-1 \le x \le 1$ $\sec^{-1}(-x) = \pi - \sec^{-1}x \;$ and $\; \tan^{-1}(-x) = -\tan^{-1}x$ for all real x. $\cot^{-1}(-x) = -\cot^{-1}x, \qquad \text{cosec}^{-1}(-x) = -\text{cosec}^{-1}x$	
$\sin^{-1}x + \cos^{-1}x = \frac{\pi}{2}, \; -1 \le x \le 1$ $\text{cosec}^{-1}x + \sec^{-1}x = \frac{\pi}{2}$ $\tan^{-1}x + \cot^{-1}x = \frac{\pi}{2}$	$\tan^{-1}x + \tan^{-1}y = \tan^{-1}\frac{x+y}{1-xy}, xy < 1$ $\tan^{-1}x - \tan^{-1}y = \tan^{-1}\frac{x-y}{1+xy}, xy < 1$
$\sin^{-1}x = \cos^{-1}\sqrt{1-x^2}, \; 0 \le x \le 1 \qquad \sin^{-1}x = -\cos^{-1}\sqrt{1-x^2}, \; -1 \le x \le 0$	
If a, b and the radicals are positive, then $\sin^{-1}\frac{a}{\sqrt{a^2+b^2}} = \cos^{-1}\frac{b}{\sqrt{a^2+b^2}} = \tan^{-1}\frac{a}{b}$	

UNIT 7

7.1 Properties of triangles

1. **Sine Rule**

In any triangle ABC, $\dfrac{a}{\sin A} = \dfrac{b}{\sin B} = \dfrac{c}{\sin C}$

2. **Cosine Rule**

In any triangle ABC, $\cos A = \dfrac{b^2+c^2-a^2}{2bc}$, $\cos B = \dfrac{c^2+a^2-b^2}{2ac}$,

$\cos C = \dfrac{a^2+b^2-c^2}{2ab}$

3. **Projection Rule :**

In any triangle ABC, $a = b\cos C + c\cos B$, $b = c\cos A + a\cos C$,

$c = a\cos B + b\cos A$

4. **Napier's Rule (tangent rule) :** In any triangle ABC,

$\tan\dfrac{B-C}{2} = \dfrac{b-c}{b+c}\cot\dfrac{A}{2}$, $\qquad \tan\dfrac{C-A}{2} = \dfrac{c-a}{c+a}\cot\dfrac{B}{2}$, $\qquad \tan\dfrac{A-B}{2} = \dfrac{a-b}{a+b}\cot\dfrac{C}{2}$

5. **Ratio of Half-Angles**

a) $\sin\dfrac{A}{2} = \sqrt{\dfrac{(s-b)(s-c)}{bc}}$, $\qquad \sin\dfrac{B}{2} = \sqrt{\dfrac{(s-c)(s-a)}{ca}}$ $\qquad \sin\dfrac{C}{2} = \sqrt{\dfrac{(s-a)(s-b)}{ab}}$

b) $\cos\dfrac{A}{2} = \sqrt{\dfrac{s(s-a)}{bc}}$ $\qquad \cos\dfrac{B}{2} = \sqrt{\dfrac{s(s-b)}{ca}}$ $\qquad \cos\dfrac{C}{2} = \sqrt{\dfrac{s(s-c)}{ab}}$

c) $\tan\dfrac{A}{2} = \sqrt{\dfrac{(s-b)(s-c)}{s(s-a)}}$, $\qquad \tan\dfrac{B}{2} = \sqrt{\dfrac{(s-c)(s-a)}{s(s-b)}}$ $\qquad \tan\dfrac{C}{2} = \sqrt{\dfrac{(s-a)(s-b)}{s(s-c)}}$

6. **Area of $\triangle$ ABC**

a) $\Delta = \dfrac{1}{2} bc \sin A = \dfrac{1}{2} ca \sin B = \dfrac{1}{2} ab \sin C$

b) $\Delta = \sqrt{s(s-a)(s-b)(s-c)}$ where $s = \dfrac{a+b+c}{2}$

c) $\Delta = \dfrac{a^2 \sin B \sin C}{2 \sin A} = \dfrac{b^2 \sin C \sin A}{2 \sin B} = \dfrac{c^2 \sin A \sin B}{2 \sin C}$

Examples 7.1

1. $\Delta = \dfrac{abc}{4R}$

$\Delta = \dfrac{1}{2} ab \sin C = \dfrac{1}{2} ab \dfrac{c}{2R}$ (since $\dfrac{c}{\sin C} = 2R$)

$= \dfrac{abc}{4R}$

2. $\Delta = 2R^2 \sin A \sin B \sin C$

$\Delta = \dfrac{1}{2} ab \sin C = \dfrac{1}{2} \cdot 2R \sin A \cdot 2R \sin B \sin C$ (since $\dfrac{a}{\sin A} = \dfrac{b}{\sin B} = 2R$)

$= 2R^2 \sin A \sin B \sin C$

3. $\Delta = \sqrt{s(s-a)(s-b)(s-c)}$

$\Delta = \dfrac{1}{2} ab \sin C = \dfrac{1}{2} ab \cdot 2 \sin \dfrac{C}{2} \cdot \cos \dfrac{C}{2}$

$= ab \sqrt{\dfrac{(s-a)(s-b)}{ab}} \cdot \sqrt{\dfrac{s(s-c)}{ab}} = \sqrt{s(s-a)(s-b)(s-c)}$

4. $\dfrac{\cos A}{a} + \dfrac{\cos B}{b} + \dfrac{\cos C}{c} = \dfrac{a^2+b^2+c^2}{2abc}$

$\cos A = \dfrac{b^2+c^2-a^2}{2bc} \Rightarrow \dfrac{\cos A}{a} = \dfrac{b^2+c^2-a^2}{2abc}$

Similarly, $\dfrac{\cos B}{b} = \dfrac{c^2+a^2-b^2}{2abc}$ and $\dfrac{\cos C}{c} = \dfrac{a^2+b^2-c^2}{2abc}$

Hence, $\dfrac{\cos A}{a} + \dfrac{\cos B}{b} + \dfrac{\cos C}{c} = \dfrac{a^2+b^2+c^2}{2abc}$

5. In any triangle ABC, $\sin\frac{B+C}{2} = \cos\frac{A}{2}$

For, $A + B + C = 180°$ => $\frac{B+C}{2} = 90° - \frac{A}{2}$

$$\sin\frac{B+C}{2} = \sin\left(90° - \frac{A}{2}\right) = \cos\frac{A}{2}$$

Exercise 7.1 : Choose the correct answer

1. In a $\triangle ABC$, $\sin\frac{A+B}{2} = $ ....

  a) $\sin\frac{C}{2}$ b) $\cos\frac{C}{2}$ c) $\tan\frac{C}{2}$ d) $\text{cosec}\frac{C}{2}$

2. In a $\triangle ABC$, $\cos\frac{A}{2} = $ ....

  a) $\sqrt{\frac{(s-b)(s-c)}{s(s-a)}}$ b) $\sqrt{\frac{s(s-a)}{(s-b)(s-c)}}$ c) $\sqrt{\frac{s(s-a)}{bc}}$ d) $\sqrt{\frac{(s-b)(s-c)}{bc}}$

3. If $a = 13$, $b = 14$, $c = 15$ then $\tan\frac{B}{2} = $ ....

  a) $\frac{3}{2}$ b) $\frac{7}{2}$ c) $\frac{4}{7}$ d) $\frac{5}{7}$

4. If $\frac{a}{\sin A} = \frac{b}{\sin B} = \frac{c}{\sin C} = K$,

  then $(b-c)\sin A + (c-a)\sin B + (a-b)\sin C = $ ....

  a) 1 b) 0 c) abc d) 2

5. If $a = 8$, $b = 9$, $c = 12$ then $\cos C = $ ....

  a) $\frac{1}{142}$ b) $\frac{1}{146}$ c) $\frac{1}{148}$ d) $\frac{1}{144}$

Answers : 1 - b 2 - c 3 - c 4 - b 5 - d

Solution 7.1

1. $\sin\frac{A+B}{2} = \sin\left(90 - \frac{C}{2}\right) = \cos\frac{C}{2}$

2. $\cos \frac{A}{2} = \sqrt{\frac{s(s-a)}{bc}}$

3. $a = 13, b = 14, c = 15$, $s = \frac{a+b+c}{2} = \frac{42}{2} = 21$

   $\tan \frac{B}{2} = \sqrt{\frac{(s-c)(s-a)}{s(s-b)}} = \sqrt{\frac{6 \times 8}{21 \times 7}} = \frac{4}{7}$

4. $\frac{a}{\sin A} = \frac{b}{\sin B} = \frac{c}{\sin C} = K$, then $\sin A = \frac{a}{K}$, $\sin B = \frac{b}{K}$, $\sin C = \frac{c}{K}$

   $(b - c) \sin A + (c - a) \sin B + (a - b) \sin C$

   $= (b - c) \frac{a}{K} + (c - a) \frac{b}{K} + (a - b) \frac{c}{K} = 0$

5. $a = 8, b = 9, c = 12$, $\cos C = \frac{a^2 + b^2 - c^2}{2ab} = \frac{8^2 + 9^2 - 12^2}{144} = \frac{1}{144}$

Table 7.1 ☺ **Try to recollect the following Table**

$\frac{a}{\sin A} = \frac{b}{\sin B} = \frac{c}{\sin C}$; $\cos A = \frac{b^2 + c^2 - a^2}{2bc}$, $\cos B = \frac{c^2 + a^2 - b^2}{2ac}$, $\cos C = \frac{a^2 + b^2 - c^2}{2ab}$

$a = b \cos C + c \cos B$, $b = c \cos A + a \cos C$, $c = a \cos B + b \cos A$

$\tan \frac{B-C}{2} = \frac{b-c}{b+c} \cot \frac{A}{2}$, $\tan \frac{C-A}{2} = \frac{c-a}{c+a} \cot \frac{B}{2}$, $\tan \frac{A-B}{2} = \frac{a-b}{a+b} \cot \frac{C}{2}$

a) $\sin \frac{A}{2} = \sqrt{\frac{(s-b)(s-c)}{bc}}$, $\sin \frac{B}{2} = \sqrt{\frac{(s-c)(s-a)}{ca}}$, $\sin \frac{C}{2} = \sqrt{\frac{(s-a)(s-b)}{ab}}$

b) $\cos \frac{A}{2} = \sqrt{\frac{s(s-a)}{bc}}$, $\cos \frac{B}{2} = \sqrt{\frac{s(s-b)}{ca}}$, $\cos \frac{C}{2} = \sqrt{\frac{s(s-c)}{ab}}$

c) $\tan \frac{A}{2} = \sqrt{\frac{(s-b)(s-c)}{s(s-a)}}$, $\tan \frac{B}{2} = \sqrt{\frac{(s-c)(s-a)}{s(s-b)}}$, $\tan \frac{C}{2} = \sqrt{\frac{(s-a)(s-b)}{s(s-c)}}$

a) $\Delta = \frac{1}{2} bc \sin A = \frac{1}{2} ac \sin B = \frac{1}{2} ab \sin C$

b) $\Delta = \sqrt{s(s-a)(s-b)(s-c)}$ where $s = \frac{a+b+c}{2}$

c) $\Delta = \frac{a^2 \sin B \sin C}{2 \sin A} = = \frac{b^2 \sin C \sin A}{2 \sin B} = = \frac{c^2 \sin A \sin B}{2 \sin C}$

7.2 Solutions of triangles

1. **Circum-Radius** : R

i) $2R = \dfrac{a}{\sin A} = \dfrac{b}{\sin B} = \dfrac{c}{\sin C}$ ii) $R = \dfrac{abc}{4\Delta}$

2. **In – Radius** : r

i) $r = \dfrac{\Delta}{s}$ ii) $r = 4R \sin \dfrac{A}{2} \sin \dfrac{B}{2} \sin \dfrac{C}{2}$

iii) $r = (s-a) \tan \dfrac{A}{2} = (s-b) \tan \dfrac{B}{2} = (s-c) \tan \dfrac{C}{2}$

3. **Ex – Radius** : r_1, r_2, r_3

i) $r_1 = \dfrac{\Delta}{s-a}, \; r_2 = \dfrac{\Delta}{s-b}, \; r_3 = \dfrac{\Delta}{s-c}$

ii) $r_1 = s \tan \dfrac{A}{2}, \; r_2 = s \tan \dfrac{B}{2}, \; r_3 = s \tan \dfrac{C}{2}$

iii) $r_1 = 4R \sin \dfrac{A}{2} \cos \dfrac{B}{2} \cos \dfrac{C}{2},$

$r_2 = 4R \cos \dfrac{A}{2} \sin \dfrac{B}{2} \cos \dfrac{C}{2},$

$r_3 = 4R \cos \dfrac{A}{2} \cos \dfrac{B}{2} \sin \dfrac{C}{2}$

Examples 7.2

1. $r = (s-a) \tan \dfrac{A}{2}$

R.H.S $= (s-a) \tan \dfrac{A}{2}$

$= (s-a) \sqrt{\dfrac{(s-b)(s-c)}{s(s-a)}}$

$= \dfrac{1}{s} \sqrt{s(s-a)(s-b)(s-c)} = \dfrac{\Delta}{s} = r$

2. $r_3 = s \tan \dfrac{C}{2}$

$$\text{L.H.S} = \frac{\Delta}{s-c} = \frac{\sqrt{s(s-a)(s-b)(s-c)}}{s-c}$$

$$= s\sqrt{\frac{(s-a)(s-b)}{s(s-c)}} = s \tan \frac{C}{2}$$

3. $r r_2 \cot \frac{B}{2} = \Delta$

$$\text{L.H.S} = r r_2 \cot \frac{B}{2} = r \cdot \frac{\Delta}{s-b} \cdot \frac{s-b}{r} = \Delta$$

4. $\dfrac{r r_1}{r_2 r_3} = \tan^2 \dfrac{A}{2}$

$$\text{L.H.S} = \frac{r r_1}{r_2 r_3} = \frac{\Delta \cdot \Delta (s-b)(s-c)}{s(s-a) \cdot \Delta \cdot \Delta}$$

$$= \frac{(s-b)(s-c)}{s(s-a)} = \tan^2 \frac{A}{2}$$

5. $r r_1 r_2 r_3 = \dfrac{\Delta}{s} \cdot \dfrac{\Delta}{s-a} \cdot \dfrac{\Delta}{s-b} \cdot \dfrac{\Delta}{s-c}$

$$= \frac{\Delta^4}{s(s-a)(s-b)(s-c)}$$

$$= \frac{\Delta^4}{\Delta^2} = \Delta^2$$

Exercise 7.2 : Choose the correct answer

1. $r^3 \cdot \cot^2 \dfrac{A}{2} \cdot \cot^2 \dfrac{B}{2} \cdot \cot^2 \dfrac{C}{2} = $ ....

a) $r r_2 r_3$ b) $r_1 r_2 r_3$ c) $r r_1 r_2$ d) $r r_1 r_3$

2. $r_1 (s-a) + r_2 (s-b) + r_3 (s-c) = $ ....

a) 3Δ b) Δ^2 c) 2Δ d) $3\Delta^2$

3. $\dfrac{bc \cos A}{2\Delta} = $ ....

a) sin A b) cos A c) cot A d) tan A

4. $4\Delta \cot A =$ ....

a) $b^2 + c^2 + a^2$ b) $a^2 + b^2 - c^2$ c) $c^2 + b^2 - a^2$ d) $b^2 + c^2 - a^2$

5. $r\, r_1 \cot \dfrac{A}{2} =$ ....

a) Δ b) Δ^2 c) 2Δ d) $2\Delta^2$

6. $r_2\, r_3 \tan \dfrac{A}{2} =$ ....

a) Δ b) Δ^2 c) 2Δ d) $2\Delta^2$

7. $\Delta \left[\cot \dfrac{A}{2} + \cot \dfrac{B}{2} + \cot \dfrac{C}{2} \right] =$ ....

a) $3s$ b) s^3 c) s^2 d) $2s$

8. If $b^2 + c^2 - a^2 = bc$, then the angle A = ...

a) $\dfrac{\pi}{6}$ b) $\dfrac{\pi}{3}$ c) $\dfrac{\pi}{4}$ d) $\dfrac{\pi}{2}$

9. $a^2 b^2 c^2 \sin A \sin B \sin C =$ ....

a) 8Δ b) $8\Delta^3$ c) $2\Delta^3$ d) $2\Delta^2$

10. $\dfrac{1}{r_1} + \dfrac{1}{r_2} + \dfrac{1}{r_3} - \dfrac{1}{r} =$ ....

a) 1 b) 2 c) 0 d) -1

Answers : 1 - b 2 - a 3 - c 4 - d 5 - a 6 - a 7 - c 8 - b 9 - b 10 - c

Solution 7.2

1. $r^3 . \cot^2 \dfrac{A}{2} . \cot^2 \dfrac{B}{2} . \cot^2 \dfrac{C}{2} = r_1\, r_2\, r_3$

L.H.S $= r^3 . \cot^2 \dfrac{A}{2} . \cot^2 \dfrac{B}{2} . \cot^2 \dfrac{B}{2}$

$= r^3 . \dfrac{(s-a)^2}{r^2} . \dfrac{(s-b)^2}{r^2} . \dfrac{(s-c)^2}{r^2}$

$$= \frac{(s-a)^2 (s-b)^2 (s-c)^2}{r^3}$$

$$= \frac{\Delta^4}{s^2 r^3} = \frac{\Delta^2}{s^2} \frac{\Delta^2}{r^3}$$

$$= r^2 \cdot \frac{\Delta^2}{r^3} = \frac{\Delta^2}{r} = r_1 r_2 r_3 \quad \text{(using example 5)}$$

2. $r_1 (s - a) + r_2 (s - b) + r_3 (s - c)$

$$= \frac{\Delta}{s-a}(s-a) + \frac{\Delta}{s-b}(s-b) + \frac{\Delta}{s-c}(s-c) = 3\Delta$$

3. $\dfrac{bc \cos A}{2\Delta} = \dfrac{bc \cos A}{bc \sin A} = \cot A \quad \left(\text{using } \Delta = \dfrac{1}{2} bc \sin A \right)$

4. $4 \Delta \cot A = 4\Delta \dfrac{\cos A}{\sin A} = 4\Delta \dfrac{b^2+c^2-a^2}{2bc} \cdot \dfrac{bc}{2\Delta} = b^2 + c^2 - a^2$

5. $r\, r_1 \cot \dfrac{A}{2} = r \dfrac{\Delta}{s-a} \dfrac{s-a}{r} = \Delta$

6. $r_2\, r_3 \tan \dfrac{A}{2} = r_2\, r_3 \dfrac{r_1}{s}$

$$= \frac{1}{s} \frac{\Delta}{s-b} \cdot \frac{\Delta}{s-c} \cdot \frac{\Delta}{s-a} = \frac{\Delta^3}{\Delta^2} = \Delta$$

7. $\Delta \left[\cot \dfrac{A}{2} + \cot \dfrac{B}{2} + \cot \dfrac{B}{2} \right] = \Delta \left[\dfrac{s-a}{r} + \dfrac{s-b}{r} + \dfrac{s-c}{r} \right]$

$$= \frac{\Delta}{r} [3s - (a+b+c)]$$

$$= \frac{\Delta}{r}(3s - 2s) = s \cdot s = s^2$$

8. $b^2 + c^2 - a^2 = bc \implies 2 bc \cos A = bc \implies \cos A = \dfrac{1}{2} \implies A = \dfrac{\pi}{3}$

9. $a^2 b^2 c^2 \sin A \sin B \sin C = a^2 b^2 c^2 \cdot \dfrac{2\Delta}{bc} \cdot \dfrac{2\Delta}{ca} \cdot \dfrac{2\Delta}{ab} = 8\Delta^3$

10. $\dfrac{1}{r_1} + \dfrac{1}{r_2} + \dfrac{1}{r_3} - \dfrac{1}{r} = \dfrac{s-a}{\Delta} + \dfrac{s-b}{\Delta} + \dfrac{s-c}{\Delta} - \dfrac{s}{\Delta}$

$$= \frac{2s-(a+b+c)}{\Delta} = \frac{2s-2s}{\Delta} = 0$$

Table 7.2  **Try to recollect the following :**

$\frac{a}{\sin A} = \frac{b}{\sin B} = \frac{c}{\sin C}$; $\cos A = \frac{b^2+c^2-a^2}{2bc}$, $\cos B = \frac{c^2+a^2-b^2}{2ac}$, $\cos C = \frac{a^2+b^2-c^2}{2ab}$
$a = b\cos C + c\cos B$, $b = c\cos A + a\cos C$, $c = a\cos B + b\cos A$
$\tan\frac{B-C}{2} = \frac{b-c}{b+c}\cot\frac{A}{2}$, $\tan\frac{C-A}{2} = \frac{c-a}{c+a}\cot\frac{B}{2}$, $\tan\frac{A-B}{2} = \frac{a-b}{a+b}\cot\frac{C}{2}$
a) $\sin\frac{A}{2} = \sqrt{\frac{(s-b)(s-c)}{bc}}$, $\sin\frac{B}{2} = \sqrt{\frac{(s-c)(s-a)}{ca}}$, $\sin\frac{C}{2} = \sqrt{\frac{(s-a)(s-b)}{ab}}$
b) $\cos\frac{A}{2} = \sqrt{\frac{s(s-a)}{bc}}$, $\cos\frac{B}{2} = \sqrt{\frac{s(s-b)}{ca}}$, $\cos\frac{C}{2} = \sqrt{\frac{s(s-c)}{ab}}$
c) $\tan\frac{A}{2} = \sqrt{\frac{(s-b)(s-c)}{s(s-a)}}$, $\tan\frac{B}{2} = \sqrt{\frac{(s-c)(s-a)}{s(s-b)}}$, $\tan\frac{C}{2} = \sqrt{\frac{(s-a)(s-b)}{s(s-c)}}$
a) $\Delta = \frac{1}{2}bc\sin A = \frac{1}{2}ac\sin B = \frac{1}{2}ab\sin C$
b) $\Delta = \sqrt{s(s-a)(s-b)(s-c)}$ where $s = \frac{a+b+c}{2}$
c) $\Delta = \frac{a^2\sin B\sin C}{2\sin A} = = \frac{b^2\sin C\sin A}{2\sin B} = = \frac{c^2\sin A\sin B}{2\sin C}$
$2R = \frac{a}{\sin A} = \frac{b}{\sin B} = \frac{c}{\sin C}$, $R = \frac{abc}{4\Delta}$, $r = \frac{\Delta}{s}$,
$r = 4R\sin\frac{A}{2}\sin\frac{B}{2}\sin\frac{C}{2}$
$r = (s-a)\tan\frac{A}{2} = (s-b)\tan\frac{B}{2} = (s-c)\tan\frac{C}{2}$
$r_1 = \frac{\Delta}{s-a}$, $r_2 = \frac{\Delta}{s-b}$, $r_3 = \frac{\Delta}{s-c}$
$r_1 = s\tan\frac{A}{2}$, $r_2 = s\tan\frac{B}{2}$, $r_3 = s\tan\frac{C}{2}$
$r_1 = 4R\sin\frac{A}{2}\cos\frac{B}{2}\cos\frac{C}{2}$, $r_2 = 4R\cos\frac{A}{2}\sin\frac{B}{2}\cos\frac{C}{2}$, $r_3 = 4R\cos\frac{A}{2}\cos\frac{B}{2}\sin\frac{C}{2}$

☺ **Try to recollect the following : Table II**

The principal value of sine function lies in $\left[-\frac{\pi}{2}, \frac{\pi}{2}\right]$, cosine function lies in $[0, \pi]$ and tan lies in $\left(-\frac{\pi}{2}, \frac{\pi}{2}\right)$	
$\sin\theta = 0 \Rightarrow \theta = n\pi, n \in Z$	$\sin\theta = \sin\alpha \Rightarrow \theta = n\pi + (-1)^n \alpha, n \in Z$
$\cos\theta = 0 \Rightarrow \theta = (2n+1)\frac{\pi}{2}, n \in Z$	$\cos\theta = \cos\alpha \Rightarrow \theta = 2n\pi \pm \alpha, \quad n \in Z$
$\tan\theta = 0 \Rightarrow \theta = n\pi, n \in Z$	$\tan\theta = \tan\alpha \Rightarrow \theta = n\pi + \alpha, \quad n \in Z$
If $a\cos\theta + b\sin\theta = c, |c| \leq \sqrt{a^2 + b^2}$, then $\theta = \alpha + 2n\pi \pm \cos^{-1}\frac{c}{r}$ where $r = \sqrt{a^2 + b^2}$ and $\alpha = \tan^{-1}\frac{b}{a}$	

$\sin^{-1}(\sin x) = x$	$\cos^{-1}(\cos x) = x$	$\tan^{-1}(\tan x) = x$
$\text{cosec}^{-1}(\text{cosec } x) = x$	$\sec^{-1}(\sec x) = x$	$\cot^{-1}(\cot x) = x$
$\sin^{-1}\frac{1}{x} = \text{cosec}^{-1} x$	$\cos^{-1}\frac{1}{x} = \sec^{-1} x$	$\tan^{-1}\frac{1}{x} = \cot^{-1} x$
$\text{cosec}^{-1}\frac{1}{x} = \sin^{-1} x$	$\sec^{-1}\frac{1}{x} = \cos^{-1} x$	$\cot^{-1}\frac{1}{x} = \tan^{-1} x$

$\sin^{-1}(-x) = -\sin^{-1} x$ and $\cos^{-1}(-x) = \pi - \cos^{-1} x$ for $-1 \leq x \leq 1$
$\sec^{-1}(-x) = \pi - \sec^{-1} x$ and $\tan^{-1}(-x) = -\tan^{-1} x$ for all real x.
$\cot^{-1}(-x) = -\cot^{-1} x, \quad \text{cosec}^{-1}(-x) = -\text{cosec}^{-1} x$

$\sin^{-1} x + \cos^{-1} x = \frac{\pi}{2},\ -1 \leq x \leq 1$	$\tan^{-1} x + \tan^{-1} y = \tan^{-1}\frac{x+y}{1-xy}, xy < 1$
$\text{cosec}^{-1} x + \sec^{-1} x = \frac{\pi}{2}$	$\tan^{-1} x - \tan^{-1} y = \tan^{-1}\frac{x-y}{1+xy}, xy < 1$
$\tan^{-1} x + \cot^{-1} x = \frac{\pi}{2}$	

$\sin^{-1} x = \cos^{-1}\sqrt{1-x^2},\ 0 \leq x \leq 1 \quad \sin^{-1} x = -\cos^{-1}\sqrt{1-x^2},\ -1 \leq x \leq 0$	
If a, b and the radicals are positive, then $\sin^{-1}\frac{a}{\sqrt{a^2+b^2}} = \cos^{-1}\frac{b}{\sqrt{a^2+b^2}} = \tan^{-1}\frac{a}{b}$	

$$\frac{a}{\sin A} = \frac{b}{\sin B} = \frac{c}{\sin C} \ ; \ \cos A = \frac{b^2+c^2-a^2}{2bc}, \ \cos B = \frac{c^2+a^2-b^2}{2ac}, \ \cos C = \frac{a^2+b^2-c^2}{2ab}$$

$$a = b\cos C + c\cos B, \ b = c\cos A + a\cos C, \ c = a\cos B + b\cos A$$

$$\tan\frac{B-C}{2} = \frac{b-c}{b+c}\cot\frac{A}{2}, \quad \tan\frac{C-A}{2} = \frac{c-a}{c+a}\cot\frac{B}{2}, \quad \tan\frac{A-B}{2} = \frac{a-b}{a+b}\cot\frac{C}{2}$$

a) $\sin\frac{A}{2} = \sqrt{\frac{(s-b)(s-c)}{bc}}, \quad \sin\frac{B}{2} = \sqrt{\frac{(s-c)(s-a)}{ca}} \quad \sin\frac{C}{2} = \sqrt{\frac{(s-a)(s-b)}{ab}}$

b) $\cos\frac{A}{2} = \sqrt{\frac{s(s-a)}{bc}} \quad \cos\frac{B}{2} = \sqrt{\frac{s(s-b)}{ca}} \quad \cos\frac{C}{2} = \sqrt{\frac{s(s-c)}{ab}}$

c) $\tan\frac{A}{2} = \sqrt{\frac{(s-b)(s-c)}{s(s-a)}}, \quad \tan\frac{B}{2} = \sqrt{\frac{(s-c)(s-a)}{s(s-b)}} \quad \tan\frac{C}{2} = \sqrt{\frac{(s-a)(s-b)}{s(s-c)}}$

a) $\Delta = \frac{1}{2} bc \sin A = \frac{1}{2} ac \sin B = \frac{1}{2} ab \sin C$

b) $\Delta = \sqrt{s(s-a)(s-b)(s-c)}$ where $s = \frac{a+b+c}{2}$

c) $\Delta = \frac{a^2 \sin B \sin C}{2 \sin A} = \frac{b^2 \sin C \sin A}{2 \sin B} = \frac{c^2 \sin A \sin B}{2 \sin C}$

$$2R = \frac{a}{\sin A} = \frac{b}{\sin B} = \frac{c}{\sin C}, \quad R = \frac{abc}{4\Delta}, \quad r = \frac{\Delta}{s},$$

$$r = 4R \sin\frac{A}{2} \sin\frac{B}{2} \sin\frac{C}{2}$$

$$r = (s-a)\tan\frac{A}{2} = (s-b)\tan\frac{B}{2} = (s-c)\tan\frac{C}{2}$$

$$r_1 = \frac{\Delta}{s-a}, \ r_2 = \frac{\Delta}{s-b}, \ r_3 = \frac{\Delta}{s-c}$$

$$r_1 = s\tan\frac{A}{2}, \ r_2 = s\tan\frac{B}{2}, \ r_3 = s\tan\frac{C}{2}$$

$$r_1 = 4R \sin\frac{A}{2} \cos\frac{B}{2} \cos\frac{C}{2}, \ r_2 = 4R \cos\frac{A}{2} \sin\frac{B}{2} \cos\frac{C}{2},$$

$$r_3 = 4R \cos\frac{A}{2} \cos\frac{B}{2} \sin\frac{C}{2}$$

Note : We are at the final stage. Now we combine the two tables together and try to recollect all the formulae so far we have practiced.

☺ **Try to recollect the following : Table I & II combined**

sin θ cosec θ cos θ sec θ tan θ cot θ	$\frac{1}{\sin θ} = \text{cosec } θ$ $\frac{1}{\sec θ} = \cos θ$	$\frac{1}{\text{cosec } θ} = \sin θ$ $\frac{1}{\tan θ} = \cot θ$	$\frac{1}{\cos θ} = \sec θ$ $\frac{1}{\cot θ} = \tan θ$	$\tan θ = \frac{\sin θ}{\cos θ}$ $\cot θ = \frac{\cos θ}{\sin θ}$

$\sin^2 θ + \cos^2 θ = 1$; $\sin^2 θ = 1 - \cos^2 θ$; $\cos^2 θ = 1 - \sin^2 θ$

$1 + \tan^2 θ = \sec^2 θ$; $\sec^2 θ - 1 = \tan^2 θ$

$1 + \cot^2 θ = \text{cosec}^2 θ$; $\text{cosec}^2 θ - 1 = \cot^2 θ$

$\sin(A + B) = \sin A \cos B + \cos A \sin B$	$\sin(A - B) = \sin A \cos B - \cos A \sin B$
$\cos(A + B) = \cos A \cos B - \sin A \sin B$	$\cos(A - B) = \cos A \cos B + \sin A \sin B$
$\tan(A + B) = \dfrac{\tan A + \tan B}{1 - \tan A \tan B}$	$\tan(A - B) = \dfrac{\tan A - \tan B}{1 + \tan A \tan B}$

$\sin(A + B)\sin(A - B) = \sin^2 A - \sin^2 B$
$\cos(A + B)\cos(A - B) = \cos^2 A - \sin^2 B$

$\sin 2A = \begin{cases} 2 \sin A \cos A \\ \dfrac{2 \tan A}{1 + \tan^2 A} \end{cases}$ $\sin A = \begin{cases} 2 \sin \dfrac{A}{2} \cos \dfrac{A}{2} \\ \dfrac{2 \tan \dfrac{A}{2}}{1 + \tan^2 \dfrac{A}{2}} \end{cases}$

$\cos 2A = \begin{cases} \cos^2 A - \sin^2 A \\ 2\cos^2 A - 1 \\ 1 - 2\sin^2 A \\ \dfrac{1 - \tan^2 A}{1 + \tan^2 A} \end{cases}$ $\cos A = \begin{cases} \cos^2 \dfrac{A}{2} - \sin^2 \dfrac{A}{2} \\ 2\cos^2 \dfrac{A}{2} - 1 \\ 1 - 2\sin^2 \dfrac{A}{2} \\ \dfrac{1 - \tan^2 \dfrac{A}{2}}{1 + \tan^2 \dfrac{A}{2}} \end{cases}$

$\tan 2A = \dfrac{2 \tan A}{1 - \tan^2 A}$ $\tan A = \dfrac{2 \tan \dfrac{A}{2}}{1 - \tan^2 \dfrac{A}{2}}$

$1 + \cos 2A = 2 \cos^2 A$ $1 + \cos A = 2 \cos^2 \dfrac{A}{2}$

$1 - \cos 2A = 2 \sin^2 A$ $1 - \cos A = 2 \sin^2 \dfrac{A}{2}$

$\sin 3A = 3 \sin A - 4 \sin^3 A$ $\sin 18° = \cos 72° = \dfrac{\sqrt{5} - 1}{4}$

$\cos 3A = 4 \cos^3 A - 3 \cos A$ $\cos 36° = \sin 54° = \dfrac{\sqrt{5} + 1}{4}$

$\tan 3A = \dfrac{3 \tan A - \tan^3 A}{1 - 3 \tan^2 A}$

$\sin C + \sin D = 2 \sin \dfrac{C+D}{2} \cos \dfrac{C-D}{2}$	$2 \sin A \cos B = \sin(A+B) + \sin(A-B)$
$\sin C - \sin D = 2 \cos \dfrac{C+D}{2} \sin \dfrac{C-D}{2}$	$2 \cos A \sin B = \sin(A+B) - \sin(A-B)$
$\cos C + \cos D = 2 \cos \dfrac{C+D}{2} \cos \dfrac{C-D}{2}$	$2 \cos A \cos B = \cos(A+B) + \cos(A-B)$
$\cos C - \cos D = -2 \sin \dfrac{C+D}{2} \cos \dfrac{C-D}{2}$	$2 \sin A \sin B = \cos(A-B) - \cos(A+B)$

The principal value of sine function lies in $\left[-\dfrac{\pi}{2}, \dfrac{\pi}{2}\right]$, cosine function lies in $[0, \pi]$ and tan lies in $\left(-\dfrac{\pi}{2}, \dfrac{\pi}{2}\right)$

$\sin \theta = 0 \Rightarrow \theta = n\pi, n \in Z$	$\sin \theta = \sin \alpha \Rightarrow \theta = n\pi + (-1)^n \alpha;\ n \in Z$
$\cos \theta = 0 \Rightarrow \theta = (2n+1)\dfrac{\pi}{2}, n \in Z$	$\cos \theta = \cos \alpha \Rightarrow \theta = 2n\pi \pm \alpha,\quad n \in Z$
$\tan \theta = 0 \Rightarrow \theta = n\pi, n \in Z$	$\tan \theta = \tan \alpha \Rightarrow \theta = n\pi + \alpha,\quad n \in Z$

If $a \cos \theta + b \sin \theta = c, |c| \leq \sqrt{a^2+b^2}$, then $\theta = \alpha + 2n\pi \pm \cos^{-1}\dfrac{c}{r}$ where $r = \sqrt{a^2+b^2}$ and $\alpha = \tan^{-1}\dfrac{b}{a}$

$\sin^{-1}(\sin x) = x$	$\cos^{-1}(\cos x) = x$	$\tan^{-1}(\tan x) = x$
$\csc^{-1}(\csc x) = x$	$\sec^{-1}(\sec x) = x$	$\cot^{-1}(\cot x) = x$
$\sin^{-1}\dfrac{1}{x} = \csc^{-1} x$	$\cos^{-1}\dfrac{1}{x} = \sec^{-1} x$	$\tan^{-1}\dfrac{1}{x} = \cot^{-1} x$
$\csc^{-1}\dfrac{1}{x} = \sin^{-1} x$	$\sec^{-1}\dfrac{1}{x} = \cos^{-1} x$	$\cot^{-1}\dfrac{1}{x} = \tan^{-1} x$

$\sin^{-1}(-x) = -\sin^{-1} x$ and $\cos^{-1}(-x) = \pi - \cos^{-1} x$ for $-1 \leq x \leq 1$
$\sec^{-1}(-x) = \pi - \sec^{-1} x$ and $\tan^{-1}(-x) = -\tan^{-1} x$ for all real x.
$\cot^{-1}(-x) = -\cot^{-1} x$, $\csc^{-1}(-x) = -\csc^{-1} x$

$\sin^{-1} x + \cos^{-1} x = \dfrac{\pi}{2},\ -1 \leq x \leq 1$	$\tan^{-1} x + \tan^{-1} y = \tan^{-1}\dfrac{x+y}{1-xy},\ xy < 1$
$\csc^{-1} x + \sec^{-1} x = \dfrac{\pi}{2}$	$\tan^{-1} x - \tan^{-1} y = \tan^{-1}\dfrac{x-y}{1+xy},\ xy < 1$
$\tan^{-1} x + \cot^{-1} x = \dfrac{\pi}{2}$	

$\sin^{-1} x = \cos^{-1}\sqrt{1-x^2},\ 0 \leq x \leq 1$ $\sin^{-1} x = -\cos^{-1}\sqrt{1-x^2},\ -1 \leq x \leq 0$

If a, b and the radicals are positive, then $\sin^{-1}\dfrac{a}{\sqrt{a^2+b^2}} = \cos^{-1}\dfrac{b}{\sqrt{a^2+b^2}} = \tan^{-1}\dfrac{a}{b}$

$$\frac{a}{\sin A} = \frac{b}{\sin B} = \frac{c}{\sin C} \; ; \; \cos A = \frac{b^2+c^2-a^2}{2bc}, \; \cos B = \frac{c^2+a^2-b^2}{2ac}, \; \cos C = \frac{a^2+b^2-c^2}{2ab}$$

$$a = b\cos C + c\cos B, \; b = c\cos A + a\cos C, \; c = a\cos B + b\cos A$$

$$\tan\frac{B-C}{2} = \frac{b-c}{b+c}\cot\frac{A}{2}, \quad \tan\frac{C-A}{2} = \frac{c-a}{c+a}\cot\frac{B}{2}, \quad \tan\frac{A-B}{2} = \frac{a-b}{a+b}\cot\frac{C}{2}$$

a) $\sin\frac{A}{2} = \sqrt{\frac{(s-b)(s-c)}{bc}}, \quad \sin\frac{B}{2} = \sqrt{\frac{(s-c)(s-a)}{ca}} \quad \sin\frac{C}{2} = \sqrt{\frac{(s-a)(s-b)}{ab}}$

b) $\cos\frac{A}{2} = \sqrt{\frac{s(s-a)}{bc}} \quad \cos\frac{B}{2} = \sqrt{\frac{s(s-b)}{ca}} \quad \cos\frac{C}{2} = \sqrt{\frac{s(s-c)}{ab}}$

c) $\tan\frac{A}{2} = \sqrt{\frac{(s-b)(s-c)}{s(s-a)}}, \quad \tan\frac{B}{2} = \sqrt{\frac{(s-c)(s-a)}{s(s-b)}} \quad \tan\frac{C}{2} = \sqrt{\frac{(s-a)(s-b)}{s(s-c)}}$

a) $\Delta = \frac{1}{2} bc \sin A = \frac{1}{2} ac \sin B = \frac{1}{2} ab \sin C$

b) $\Delta = \sqrt{s(s-a)(s-b)(s-c)}$ where $s = \frac{a+b+c}{2}$

c) $\Delta = \frac{a^2 \sin B \sin C}{2 \sin A} = = \frac{b^2 \sin C \sin A}{2 \sin B} = = \frac{c^2 \sin A \sin B}{2 \sin C}$

$2R = \frac{a}{\sin A} = \frac{b}{\sin B} = \frac{c}{\sin C}, \quad R = \frac{abc}{4\Delta}, \quad r = \frac{\Delta}{s},$

$r = 4R \sin\frac{A}{2} \sin\frac{B}{2} \sin\frac{C}{2}$

$r = (s-a) \tan\frac{A}{2} = (s-b) \tan\frac{B}{2} = (s-c) \tan\frac{C}{2}$

$r_1 = \frac{\Delta}{s-a}, \; r_2 = \frac{\Delta}{s-b}, \; r_3 = \frac{\Delta}{s-c}$

$r_1 = s \tan\frac{A}{2}, \; r_2 = s \tan\frac{B}{2}, \; r_3 = s \tan\frac{C}{2}$

$r_1 = 4R \sin\frac{A}{2} \cos\frac{B}{2} \cos\frac{C}{2}, \; r_2 = 4R \cos\frac{A}{2} \sin\frac{B}{2} \cos\frac{C}{2}, \; r_3 = 4R \cos\frac{A}{2} \cos\frac{B}{2} \sin\frac{C}{2}$

Note :

At this stage, you must be able to recollect all the formulae given in the above table (I and II combined) on your screen.

This shows that we have achieved our aim .

Wish you all the best!

APPENDIX 1

How do the ratios (sine, cosine and tan only) look like?

1. **Sine curve**

i) Table

x	...	-2π	$-\dfrac{3\pi}{2}$	$-\pi$	$-\dfrac{\pi}{2}$	0	$\dfrac{\pi}{2}$	π	$\dfrac{3\pi}{2}$	2π	$\dfrac{5\pi}{2}$	......
$\sin x$	...	0	1	0	-1	0	1	0	-1	0	1	...

ii) Draw the curve using the above points

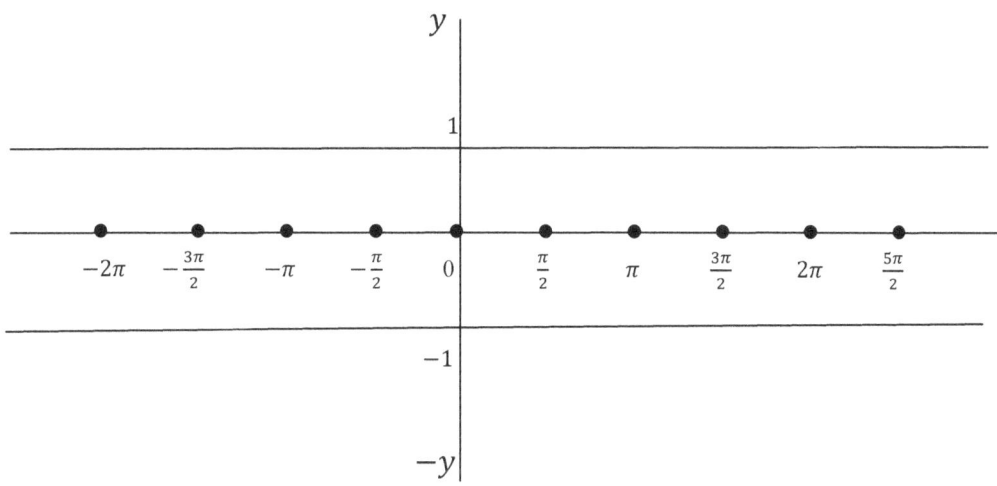

Observation

1. Sine curve is continuous from $x = -\infty$ to $x = \infty$
2. The curve lies between $y = -1$ and $y = 1$ ie $|y| = |\sin x| \leq 1 \ \forall \ x$
3. The curve cuts the x – axis at $x = n\pi$, $n = 0, \pm 1, \pm 2, \pm 3 \ldots\ldots\ldots$
4. As x increases from 0 to $\dfrac{\pi}{2}$, the curve increases from 0 to 1

   As x increases from $\dfrac{\pi}{2}$ to π, the curve decreases from 1 to 0

   As x increases from π to $\dfrac{3\pi}{2}$, the curve decreases from 0 to -1

   As x increases from $\dfrac{3\pi}{2}$ to 2π, the curve increases from -1 to 0 etc.

5. The curve has maximum value 1 at $x = \dfrac{\pi}{2}, \dfrac{5\pi}{2}, \dfrac{9\pi}{2} \ldots$

and $x = -\frac{3\pi}{2}, -\frac{7\pi}{2}, -\frac{11\pi}{2} \ldots$

The curve has minimum value -1 at $x = \frac{3\pi}{2}, \frac{7\pi}{2}, \frac{11\pi}{2} \ldots$

and $x = -\frac{\pi}{2}, -\frac{5\pi}{2}, -\frac{9\pi}{2} \ldots$

6. It is a periodic function with period 2π

2. Cosine curve

i) Table

x	...	-2π	$-\frac{3\pi}{2}$	$-\pi$	$-\frac{\pi}{2}$	0	$\frac{\pi}{2}$	π	$\frac{3\pi}{2}$	2π	$\frac{5\pi}{2}$	......
$\cos x$	...	1	0	-1	0	1	0	-1	0	1	0	...

ii) Draw the curve using the above points

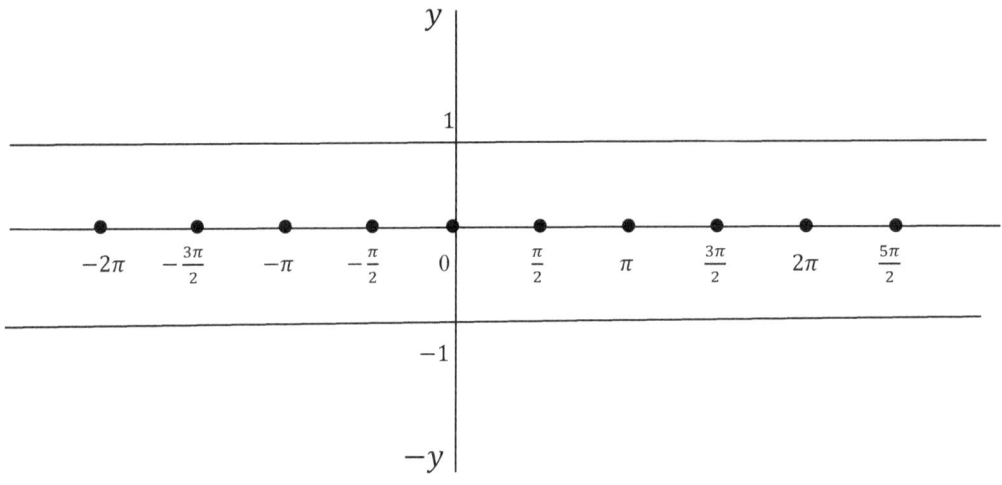

Observation

1. Cosine curve is continuous from $x = -\infty$ to $x = \infty$

2. The curve lies between $y = -1$ and $y = 1$ ie $|y| = |\cos x| \leq 1 \ \forall \ x$

3. The curve cuts the x – axis at $x = \pm\frac{\pi}{2}, \pm\frac{3\pi}{2}, \pm\frac{5\pi}{2} \ldots \ldots$

4. As x increases from 0 to $\frac{\pi}{2}$, the curve decreases from 1 to 0

   As x increases from $\frac{\pi}{2}$ to π, the curve decreases from 0 to -1

   As x increases from π to $\frac{3\pi}{2}$, the curve increases from -1 to 0

As x increases from $\frac{3\pi}{2}$ to 2π, the curve increases from 0 to 1 etc.

5. The curve has maximum value 1 at $x = 0, \pm 2\pi, \pm 4\pi, \ldots\ldots$

   The curve has minimum value -1 at $x = \pm \pi, \pm 3\pi, \ldots\ldots\ldots$

6. It is a periodic function with period 2π

3. Tangent curve

i) Table

x	…………	0 to $\frac{\pi}{2}$	$\frac{\pi}{2}$ to π	π to $\frac{3\pi}{2}$	…………
$\tan x$ …………		$\tan x$ + ve 0 to ∞	$\tan x$ - ve $-\infty$ to 0	$\tan x$ + ve 0 to ∞	…………

ii) Draw the curve using the above points

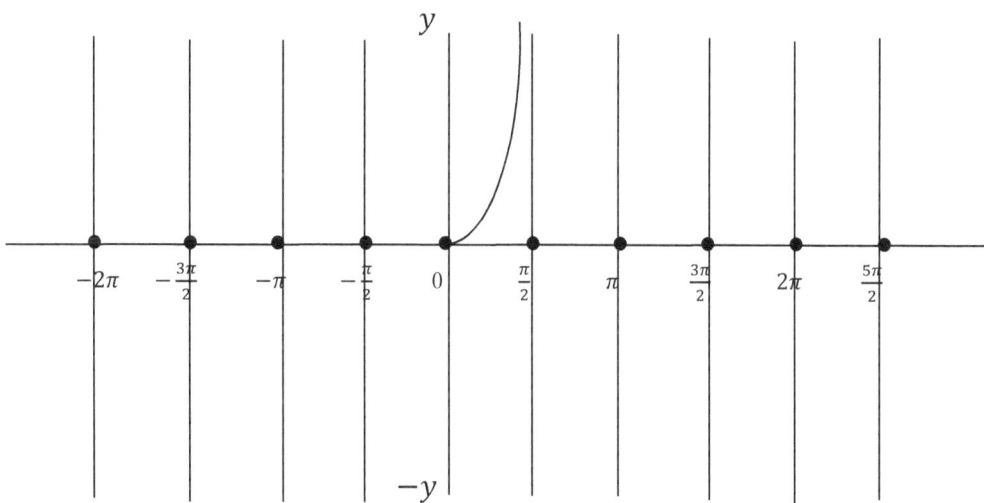

Observation

1. Tangent curve is not continuous.

   It consists of infinite number of broken pieces.

2. The curve is not defined at $x = \pm\frac{\pi}{2}, \pm\frac{3\pi}{2}, \pm\frac{5\pi}{2}\ldots\ldots$

3. The curve cuts the x – axis at $x = 0, \pm\pi, \pm 2\pi, \ldots\ldots\ldots$

4. As x increases from 0 to $\frac{\pi}{2}$, the curve increases from 0 to ∞

   As x increases from $\frac{\pi}{2}$ to π, the curve increases from $-\infty$ to 0

   As x increases from π to $\frac{3\pi}{2}$, the curve increases from 0 to ∞

   As x increases from $\frac{3\pi}{2}$ to 2π, the curve increases from $-\infty$ to 0 etc.

5. It is a periodic function with period π

Exercise : State TRUE or FALSE

1. The curve $y = \tan x$ is continuous in $-\infty \leq x \leq \infty$
2. Any straight line parallel to x – axis cuts $y = \tan x$ at one point only.
3. As x increases from 0 to $\frac{\pi}{2}$, $y = \tan x$ increases from 0 to ∞.
4. As x increases from $\frac{\pi}{2}$ to π, $y = \cos x$ decreases from 0 to -1.
5. The curve $y = \cos x$ is not defined at x = π.
6. The straight line $y = k > 1$ (parallel to x – axis) does not cut $y = \sin x$.
7. The straight line $y = 1$ touches $y = \sin x$ at infinite number of points.
8. The straight line $y = k < -1$ cuts $y = \cos x$ at infinite number of points.
9. The graph of $y = \tan x$ between $-\frac{\pi}{2}$ and $\frac{\pi}{2}$ is repeated in every interval of π on the x – axis. Hence it is a periodic function with period π.
10. The curve y = cos x lies above the line y = 1.

Answers :

1 – F 2 – F 3 – T 4 – T 5 – F 6 – T 7 – T 8 – F 9 – T 10 – F

Note : *With these basic ideas, you can easily understand the graphs of other curves given in your text book.*

Appendix II

Measurement of Angles

1. **Sexagesimal system**

    - Here the angles are measured in degrees, minutes and seconds

    [$1° = 60$ minutes ie $60'$ $1' = 60$ seconds ie $60''$]

2. **Radian** - the angle subtended at the centre of any circle by an arc whose length is equal to the radius of the circle. It is denoted by 1^c.

3. **Circular measure** : [radian measure] - the measure of any angle in terms of a radian

4. Relation between **sexagesimal** and **circular measure** :

    1 Radian = $1^c = \dfrac{180°}{\pi} = 57°17'45''$ approx. $1° = \dfrac{\pi^c}{180}$

Exercise : Choose the correct answer

1. $\dfrac{\pi}{12}$ radians = ...

    a) $25°$ b) $15°$ c) $40°$ d) $60°$

2. $150° =$ ....

    a) $\dfrac{5\pi}{6}$ b) $\dfrac{\pi}{2}$ c) $\dfrac{3\pi}{2}$ d) $\dfrac{\pi}{3}$

3. $8^c =$ ....

    a) $34°42'45''$ b) $89°24'56''$ c) $458°22'$ d) $120°$

4. $110°30' =$ ....

    a) $\dfrac{123\pi}{360}$ b) $\dfrac{213\pi}{360}$ c) $\dfrac{221\pi}{360}$ d) $\dfrac{\pi}{360}$

5. $40°15'36'' =$ ....

    a) 2.236π b) 22.36π c) 0.223π d) 2.236

Answers :   1 - b   2 - a   3 - c   4 - c   5 - c

Solution

1. $\dfrac{\pi}{12}$ radians = $\dfrac{\pi}{12} \times \dfrac{180°}{\pi} = 15°$

2. $150° = 150 \times \dfrac{\pi}{180} = \dfrac{5\pi}{6}$

3. $8^c = 8 \times \dfrac{180}{\pi} = 8 \times 180 \times \dfrac{7}{22} = 458°22'$

4. $110°30' = 110.5° = 110.5 \times \dfrac{\pi}{180} = \dfrac{221\pi}{360}$

5. $40°15'36'' = 40°15\dfrac{3}{5}' = 40.26° = 40.26 \times \dfrac{\pi}{180} = 0.2236\pi$

Appendix III

Values of trigonometric ratios in some special cases :

1. angle of **30°**   2. angle of **45°**   3. angle of **60°**

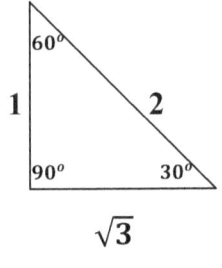 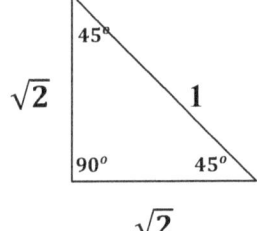 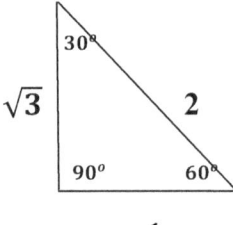

30° : 60° : 90°   45° : 45° : 90°   60° : 30° : 90°

1 : √3 : 2   **√2 : √2 : 1**   **√3 : 1 : 2**

Exercise : Complete the following

sin 30° = ….	sin 45° = …..	cosec 30° = ….
sec 45° = …..	sin 60° = ….	cosec 90° = …..
cos 30° = ….	sec 30° = …..	cosec 45° = ….
cos 45° = ….	cos 60° = ….	sec 60° = …..
tan 30° = ….	cot 30° = …..	tan 45° = ….
cot 45° = …..	tan 60° = ….	cot 90° = ….

Self Evaluation Test

1. Match the following

Column 1	Column 2
1. $\dfrac{2\tan A}{1+\tan^2 A}$	i) $\dfrac{1}{2} ab \sin C$, $\dfrac{abc}{4R}$, $s\,r$
2. $cosec^2\theta - 1$	ii) $\cos \dfrac{C}{2}$
3. $\sin^{-1}x + \cos^{-1}x$	iii) $2\cos\dfrac{C+D}{2} \cos\dfrac{C-D}{2}$
4. Δ	iv) $\theta = 2n\pi \pm \alpha \quad n \in Z, \alpha \in [0, \pi]$
5. $\sin^{-1}x$	v) $2\cos^2 \dfrac{A}{2} - 1$; $1 - 2\sin^2 \dfrac{A}{2}$; $\dfrac{1-\tan^2\frac{A}{2}}{1+\tan^2\frac{A}{2}}$
6. $\sqrt{\dfrac{s(s-c)}{ab}}$	vi) $\dfrac{\pi}{2}$, $-1 \leq x \leq 1$
7. $\cos\theta = \cos\alpha$	vii) $cosec^{-1}\dfrac{1}{x}$; $-\sin^{-1}(-x)$; $\cos^{-1}\sqrt{1-x^2}$
8. $\cos A$	viii) $\sin^{-1}\dfrac{12}{13}$
9. $\cos^{-1}\dfrac{5}{13}$	ix) $\cot^2 \theta$
10. $\cos C + \cos D$	x) $\sin 2A$

Write the answers here.

1. .....   2. .....   3. .....   4. .....   5. .....   6. .....
7. .....   8. .....   9. .....   10. .....

Answers :

1. x   2. ix   3. vi   4. i   5. vii   6. ii
7. iv   8. v   9. viii   10. iii

2. Fill in the blanks I

1. $\tan\theta = 0 \Rightarrow \theta = $ ....

2. $\cos\ldots^o = \sin\ldots^o = \dfrac{\sqrt{5}+1}{4}$

3. $\tan\theta \cdot \cos\theta = $ ....

4. $\tan(A-B) = \dfrac{\tan A - \tan B}{1 \ldots\ldots\ldots}$

5. $\cos 3A = $ .... $- 3\cos A$

6. $\sin(A+B) - \sin(A-B) = $ ………

7. The principal values of sine function lies in ………………

8. $sec^{-1}(-x) = $ …………………

9. $\cos B = \dfrac{\ldots\ldots\ldots}{2ac}$

10. $\Delta = \dfrac{c^2 \ldots\ldots\ldots}{\ldots\ldots\ldots}$

11. $1 + \ldots\ldots = \sec^2\theta$

12. $\cos\theta = 0 \Rightarrow \theta = $ ………………

13. ………… $(\sec x) = x$

14. $tan^{-1}x - tan^{-1}y = tan^{-1}\ldots\ldots\ldots$, $xy < 1$

15. As x increases from 0 to $\dfrac{\pi}{2}$, $y = \sin x$ ……… (increases/decreases) from 0 to 1

16. The curve $y = \sin x$ lies between $y = $ ……… and $y = $ …………

17. The value of the curve $y = \cos x$ at $x = -\dfrac{\pi}{2}$ is …………………

18. The period of the function $y = \cos x$ is …………

19. The curve $y = \tan x$ is ………… (defined/not defined) at $x = \dfrac{3\pi}{2}$

20. As x increases from $\dfrac{\pi}{2}$ to π, $y = \tan x$ ……… (decreases/increases) from $-\infty$ to 0

Answers :

1. $n\pi, n \in Z$ 2. $36^o, 54^o$ 3. $\sin\theta$ 4. $+\tan A \tan B$ 5. $4\cos^3 A$

6. $2\cos A \sin B$ 7. $[-\dfrac{\pi}{2}, \dfrac{\pi}{2}]$ 8. $\pi - \sec^{-1}x$ 9. $c^2 + a^2 - b^2$ 10. $\dfrac{\sin A \sin B}{2\sin C}$

11. $\tan^2\theta$ 12. $(2n+1)\dfrac{\pi}{2}, n \in Z$ 13. $\sec^{-1}$ 14. $\dfrac{x-y}{1+xy}$ 15. Increases

16. -1 to 1 17. 0 18. 2π 19. not defined 20. increases

3. Correct the following errors if any :

1. $\Delta = \sqrt{s(s-a)(s-b)(s-c)}$ where $s = \dfrac{a+b+c}{3}$

2. $cosec^{-1} x = \dfrac{\pi}{2} - sec^{-1} x$

3. $\tan \dfrac{A-B}{2} = \dfrac{a+b}{a-b} \cot \dfrac{C}{2}$

4. $\cos(45 - A) \cos(45 - B) + \sin(45 - A) \sin(45 - B) = \cos(A+B)$

5. $\sec(270 + \theta) = -cosec\,\theta$

6. $\cot(-\theta) = -\cot\theta$

7. $(1 - \sin^2\theta) \cdot \dfrac{1}{\sec^2\theta} = \cos^4\theta$

8. $\cos 30^o = \sin 45^o$

9. $\sin 1830^o = \dfrac{1}{2}$

10. The curve $y = \tan x$ is continuous in the interval $-\infty \leq x \leq \infty$.

Answers :

1. $\Delta = \sqrt{s(s-a)(s-b)(s-c)}$ where $s = \dfrac{a+b+c}{2}$

3. $\tan \dfrac{A-B}{2} = \dfrac{a-b}{a+b} \cot \dfrac{C}{2}$

5. $\sec(270 + \theta) = +cosec\,\theta$

8. $\cos 30^o = \sin 60^o$ **(or)** $\cos 45^o = \sin 45^o$

10. The curve $y = \tan x$ is not continuous in the interval $-\infty \leq x \leq \infty$.

Note : *Considering the simplicity of the questions given in the Self Evaluation Test, only answers (no solutions) are given.*

Tables

Table 1. Expressions of each ratio in terms of other ratios.

Verification of the following results is left to your convenience!

	$\sin\theta$	$\cos\theta$	$\tan\theta$	$\cot\theta$	$\sec\theta$	$\csc\theta$
$\sin\theta$	$\sin\theta$	$\sqrt{1-\cos^2\theta}$	$\dfrac{\tan\theta}{\sqrt{1+\tan^2\theta}}$	$\dfrac{1}{\sqrt{1+\cot^2\theta}}$	$\dfrac{\sqrt{\sec^2\theta-1}}{\sec\theta}$	$\dfrac{1}{\csc\theta}$
$\cos\theta$	$\sqrt{1-\sin^2\theta}$	$\cos\theta$	$\dfrac{1}{\sqrt{1+\tan^2\theta}}$	$\dfrac{\cot\theta}{\sqrt{1+\cot^2\theta}}$	$\dfrac{1}{\sec\theta}$	$\dfrac{\sqrt{\csc^2\theta-1}}{\csc\theta}$
$\tan\theta$	$\dfrac{\sin\theta}{\sqrt{1-\sin^2\theta}}$	$\dfrac{\sqrt{1-\cos^2\theta}}{\cos\theta}$	$\tan\theta$	$\dfrac{1}{\cot\theta}$	$\sqrt{\sec^2\theta-1}$	$\dfrac{1}{\sqrt{\csc^2\theta-1}}$
$\csc\theta$	$\dfrac{1}{\sin\theta}$	$\dfrac{1}{\sqrt{1-\cos^2\theta}}$	$\dfrac{\sqrt{1+\tan^2\theta}}{\tan\theta}$	$\sqrt{1+\cot^2\theta}$	$\dfrac{\sec\theta}{\sqrt{\sec^2\theta-1}}$	$\csc\theta$
$\sec\theta$	$\dfrac{1}{\sqrt{1-\sin^2\theta}}$	$\dfrac{1}{\cos\theta}$	$\sqrt{1+\tan^2\theta}$	$\dfrac{\sqrt{1+\cot^2\theta}}{\cot\theta}$	$\sec\theta$	$\dfrac{\csc\theta}{\sqrt{\csc^2\theta-1}}$
$\cot\theta$	$\dfrac{\sqrt{1-\sin^2\theta}}{\sin\theta}$	$\dfrac{\cos\theta}{\sqrt{1-\cos^2\theta}}$	$\dfrac{1}{\tan\theta}$	$\cot\theta$	$\dfrac{1}{\sqrt{\sec^2\theta-1}}$	$\sqrt{\csc^2\theta-1}$

Table 2. Trigonometric Functions

– *Domain, Range (Principal value) and Period*

Graph of the functions	Domain	Range (Principal value)	Period
$y = \sin x$	**R**	$[-1, 1]$	2π
$y = \cos x$	**R**	$[-1, 1]$	2π
$y = \tan x$	$\mathbf{R} - \{x : x = (2n+1)\frac{\pi}{2}, n \in \mathbf{Z}\}$	**R**	π
$y = \operatorname{cosec} x$	$\mathbf{R} - \{x : x = n\pi, n \in \mathbf{Z}\}$	$\mathbf{R} - (-1, 1)$	2π
$y = \sec x$	$\mathbf{R} - \{x : x = (2n+1)\frac{\pi}{2}, n \in \mathbf{Z}\}$	$\mathbf{R} - (-1, 1)$	2π
$y = \cot x$	$\mathbf{R} - \{x : x = n\pi, n \in \mathbf{Z}\}$	**R**	π

Table 3. Trigonometrical ratios of particular angles

[A very easy method to remember!]

1. In the 'sine' row, **the values of the four cells to the right side** of **1** *are in the reverse order that of the left side cells.* Isn't it?

2. In the 'cosine' row, **the values of the first five cells** *are in the reverse order that of the first five cells of the 'sine' row.* Isn't it?

3. In the 'cosine' row, **the values of the four cells to the right side** of **0** *are in the reverse order with the negative sign that of the left side cells.* Isn't it?

4. Try to complete the values of the remaining cells. **You can do it!**

	0°	30°	45°	60°	90°	120°	135°	150°	180°
sin	0	$\frac{1}{2}$	$\frac{1}{\sqrt{2}}$	$\frac{\sqrt{3}}{2}$	1	$\frac{\sqrt{3}}{2}$	$\frac{1}{\sqrt{2}}$	$\frac{1}{2}$	0
cos	1	$\frac{\sqrt{3}}{2}$	$\frac{1}{\sqrt{2}}$	$\frac{1}{2}$	0	$-\frac{1}{2}$	$-\frac{1}{\sqrt{2}}$	$-\frac{\sqrt{3}}{2}$	-1
tan									
cosec									
sec									
cot									

Table 4. Try to complete this. You can easily remember this table also!

	0°	90°	180°	270°	360°
sin	0	1	0	-1	0
cos	1	0	-1	0	1
tan					
cosec					
sec					
cot					

Table 5. Trigonometrical ratios of related angles :

Note : What is the significance of the shaded rows?

	sin	cos	tan	cosec	sec	cot
$-\theta$	$-\sin\theta$	$\cos\theta$	$-\tan\theta$	$-\csc\theta$	$\sec\theta$	$-\cot\theta$
$90-\theta$	$\cos\theta$	$\sin\theta$	$\cot\theta$	$\sec\theta$	$\csc\theta$	$\tan\theta$
$90+\theta$	$\cos\theta$	$-\sin\theta$	$-\cot\theta$	$\sec\theta$	$-\csc\theta$	$-\tan\theta$
$180-\theta$	$\sin\theta$	$-\cos\theta$	$-\tan\theta$	$\csc\theta$	$-\sec\theta$	$-\cot\theta$
$180+\theta$	$-\sin\theta$	$-\cos\theta$	$\tan\theta$	$-\csc\theta$	$-\sec\theta$	$\cot\theta$
$270-\theta$	$-\cos\theta$	$-\sin\theta$	$\cot\theta$	$-\sec\theta$	$-\csc\theta$	$\tan\theta$
$270+\theta$	$-\cos\theta$	$\sin\theta$	$-\cot\theta$	$-\sec\theta$	$\csc\theta$	$-\tan\theta$
$360-\theta$	$-\sin\theta$	$\cos\theta$	$-\tan\theta$	$-\csc\theta$	$\sec\theta$	$-\cot\theta$
$360+\theta$	$\sin\theta$	$\cos\theta$	$\tan\theta$	$\csc\theta$	$\sec\theta$	$\cot\theta$

Table 6. Inverse Trigonometrical Functions

— *Domain and Range (Principal value)*

Graph of the functions	Domain	Range (Principal value)
$y = \sin^{-1} x$	$[-1, 1]$	$[-\frac{\pi}{2}, \frac{\pi}{2}]$
$y = \cos^{-1} x$	$[-1, 1]$	$[0, \pi]$
$y = \tan^{-1} x$	R	$(-\frac{\pi}{2}, \frac{\pi}{2})$
$y = \csc^{-1} x$	$R - (-1, 1)$	$[-\frac{\pi}{2}, \frac{\pi}{2}] - \{0\}$
$y = \sec^{-1} x$	$R - (-1, 1)$	$[0, \pi] - \{\frac{\pi}{2}\}$
$y = \cot^{-1} x$	R	$(0, \pi)$

Practice! Practice!

A. 1. Which of the following is **not equal to** $\sin\theta$

a) $\sqrt{1-\cos^2\theta}$
b) $\dfrac{\tan\theta}{\sqrt{1+\tan^2\theta}}$
c) $\dfrac{\sqrt{\sec^2\theta-1}}{\sec\theta}$
d) $\dfrac{1}{\sqrt{1+\tan^2\theta}}$

2. Which of the following is **not equal to** $\cos\theta$

a) $\dfrac{1}{\sec\theta}$
b) $\dfrac{\sqrt{\csc^2\theta-1}}{\csc\theta}$
c) $\dfrac{1}{\sqrt{1+\tan^2\theta}}$
d) $\sqrt{1-\cos^2\theta}$

3. Find the odd man out

a) $\dfrac{\sqrt{1-\cos^2\theta}}{\cos\theta}$
b) $\dfrac{\sqrt{1+\tan^2\theta}}{\tan\theta}$
c) $\dfrac{\sec\theta}{\sqrt{\sec^2\theta-1}}$
d) $\csc\theta$

4. Find the odd man out

a) $\dfrac{1}{\sqrt{\csc^2\theta-1}}$
b) $\dfrac{\sqrt{1-\cos^2\theta}}{\cos\theta}$
c) $\cot\theta$
d) $\dfrac{\sin\theta}{\sqrt{1-\sin^2\theta}}$

5. Which of the following is **equal to** $\sec\theta$

a) $\dfrac{\sqrt{\csc^2\theta-1}}{\csc\theta}$
b) $\dfrac{\sqrt{1+\cot^2\theta}}{\cot\theta}$
c) $\dfrac{\sqrt{1-\cos^2\theta}}{\cos\theta}$
d) $\dfrac{1}{\sqrt{1+\tan^2\theta}}$

6. Which of the following is **equal to** $\cot\theta$

a) $\dfrac{1}{\tan\theta}$
b) $\dfrac{1}{\sqrt{1+\tan^2\theta}}$
c) $\dfrac{\sqrt{1-\cos^2\theta}}{\cos\theta}$
d) $\dfrac{\sqrt{1+\tan^2\theta}}{\tan\theta}$

7. Which of the following **is/are equal to** $\sin\theta$

a) $\dfrac{\sqrt{1+\cot^2\theta}}{\cot\theta}$
b) $\dfrac{\tan\theta}{\sqrt{1+\tan^2\theta}}$
c) $\dfrac{\sqrt{\sec^2\theta-1}}{\sec\theta}$
d) $\dfrac{1}{\sqrt{1+\tan^2\theta}}$

8. Which of the following **is/are not equal to** $\sec\theta$

a) $\dfrac{1}{\cos\theta}$
b) $\dfrac{\sqrt{1+\cot^2\theta}}{\cot\theta}$
c) $\dfrac{\sqrt{1-\cos^2\theta}}{\cos\theta}$
d) $\dfrac{1}{\sqrt{1+\tan^2\theta}}$

9. Which of the following **is/are equal to** $\tan\theta$

a) $\dfrac{1}{\sqrt{cosec^2\theta - 1}}$ b) $\dfrac{\sqrt{1-cos^2\theta}}{cos\theta}$ c) $cot\theta$ d) $\dfrac{sin\theta}{\sqrt{1-sin^2\theta}}$

10. The sum of the squares of any two of the following expressions is equal to 1. Find them.

a) $\dfrac{1}{\sqrt{1+tan^2\theta}}$ b) $\dfrac{\sqrt{1-cos^2\theta}}{cos\theta}$ c) $\dfrac{\sqrt{sec^2\theta-1}}{sec\theta}$ d) $\dfrac{sin\theta}{\sqrt{1-sin^2\theta}}$

11. Three of the following expressions form a trigonometric identity, Find them and also the identity.

a) $\dfrac{1}{cos\theta}$ b) $\dfrac{\sqrt{1+tan^2\theta}}{tan\theta}$ c) $\dfrac{\sqrt{1-cos^2\theta}}{cos\theta}$ d) 1

12. Three of the following expressions form a trigonometric identity, Find them and also the identity.

a) $\sqrt{1+cot^2\theta}$ b) 1 c) $\dfrac{\sqrt{1-cos^2\theta}}{cos\theta}$ d) $\dfrac{cos\theta}{\sqrt{1-cos^2\theta}}$

Answers : A. 1. d   2. d   3. a   4. c   5. b

6. a   7. b & c   8. c & d   9. a, b & d   10. a & c

11. a, c & d. $d + c^2 = a^2$ (or) $d = a^2 - c^2$ (or) $c^2 = a^2 - d$

12. a, b & d. $b + d^2 = a^2$ (or) $d^2 = a^2 - b$ (or) $b = a^2 - d^2$

B. 1. The domain of $sin\,x$ is

a) $\mathbf{R} - \{x : x = (2n+1)\dfrac{\pi}{2}, n \in \mathbf{Z}\}$ b) $\mathbf{R}$

c) $\mathbf{R} - \{x : x = n\pi, n \in \mathbf{Z}\}$ d) $\mathbf{R} - (-1, 1)$

2. The domain of $cos\,x$ is

a) $\mathbf{R} - \{x : x = (2n+1)\dfrac{\pi}{2}, n \in \mathbf{Z}\}$ b) $\mathbf{R}$

c) $\mathbf{R} - \{x : x = n\pi, n \in \mathbf{Z}\}$ d) $\mathbf{R} - (-1, 1)$

3. The domain of $tan\,x$ is

a) $\mathbf{R} - \{x : x = (2n+1)\frac{\pi}{2}, n \in \mathbf{Z}\}$ b) $\mathbf{R}$

c) $\mathbf{R} - \{x : x = n\pi, n \in \mathbf{Z}\}$ d) $\mathbf{R} - (-1, 1)$

4. The domain of **cot x** is

a) $\mathbf{R} - \{x : x = (2n+1)\frac{\pi}{2}, n \in \mathbf{Z}\}$ b) $\mathbf{R}$

c) $\mathbf{R} - \{x : x = n\pi, n \in \mathbf{Z}\}$ d) $\mathbf{R} - (-1, 1)$

5. The domain of **cosec x** is

a) $\mathbf{R} - \{x : x = (2n+1)\frac{\pi}{2}, n \in \mathbf{Z}\}$ b) $\mathbf{R}$

c) $\mathbf{R} - \{x : x = n\pi, n \in \mathbf{Z}\}$ d) $\mathbf{R} - (-1, 1)$

6. The domain of **sec x** is

a) $\mathbf{R} - \{x : x = (2n+1)\frac{\pi}{2}, n \in \mathbf{Z}\}$ b) $\mathbf{R}$

c) $\mathbf{R} - \{x : x = n\pi, n \in \mathbf{Z}\}$ d) $\mathbf{R} - (-1, 1)$

Answers : B.   1. b   2. b   3. a   4. c   5. c   6. a

C. 1. The range (Principal value) of **sin x** is

a) $[-1, 1]$ b) $\mathbf{R}$ c) $\mathbf{R} - [-1, 1]$ d) $\mathbf{R} - (-1, 1)$

2. The range (Principal value) of **cos x** is

a) $(-1, 1)$ b) $\mathbf{R}$ c) $[-1, 1]$ d) $\mathbf{R} - (-1, 1)$

3. The range (Principal value) of **tan x** is

a) $(-1, 1)$ b) $\mathbf{R}$ c) $[-1, 1]$ d) $\mathbf{R} - (-1, 1)$

4. The range (Principal value) of **sec x** is

a) $(-1, 1)$ b) $\mathbf{R}$ c) $\mathbf{R} - [-1, 1]$ d) $\mathbf{R} - (-1, 1)$

5. The range (Principal value) of **cosec x** is

a) $(-1, 1)$ b) $\mathbf{R}$ c) $\mathbf{R} - [-1, 1]$ d) $\mathbf{R} - (-1, 1)$

6. The range (Principal value) of **cot x** is

a) $(-1, 1)$ b) $\mathbf{R}$ c) $\mathbf{R} - [-1, 1]$ d) $\mathbf{R} - (-1, 1)$

Answers : C. 1. a   2. c   3. b   4. d   5. d   6. b

D. 1. The period of **tan x** is

a) π b) 2π c) 3π d) $\frac{\pi}{2}$

2. The period of **sin x** is

a) π b) 2π c) 3π d) $\frac{\pi}{2}$

3. The period of **cos x** is

a) π b) 3π c) 2π d) $\frac{\pi}{2}$

4. The period of **cosec x** is

a) π b) 3π c) 2π d) $\frac{\pi}{2}$

5. The period of **cot x** is

a) π b) 3π c) 2π d) $\frac{\pi}{2}$

6. The period of **sec x** is

a) π b) 3π c) 2π d) $\frac{\pi}{2}$

Answers : D. 1. a   2. b   3. c   4. c   5. a   6. c

E. 1. For $90 < \theta < 180$, θ lies in the .... quadrant.

a) I   b) II   c) III   d) IV

2. For $0 < \theta < 90$, $270 + \theta$ lies in the .... quadrant.

a) I   b) II   c) III   d) IV

3. $\theta = \frac{7\pi}{3}$ lies in the .... quadrant.

a) I   b) II   c) III   d) IV

4. The angles which lie in the **third** quadrant are

a) $90 - \theta, 360 + \theta$ b) $270 - \theta, 180 + \theta$

c) $270 + \theta, 360 - \theta$ d) $-\theta, 360 - \theta$

5. The angles which lie in the **fourth** quadrant are

a) $90 - \theta, 360 + \theta$ b) $270 - \theta, 180 + \theta$

c) $90 + \theta, 180 - \theta$ d) $-\theta, 270 + \theta, 360 - \theta$

6. In the **second quadrant** which of the ratios are **positive**? $90 < \theta < 180$

a) All ratios
b) $\sin \theta$, $\csc \theta$
c) $\tan \theta$, $\cot \theta$
d) $\cos \theta$, $\sec \theta$

7. For **which values** of the ratios there is **no change in ratio**?

a) $90 \pm \theta$, $360 \pm \theta$
b) $270 \pm \theta$, $180 + \theta$
c) $270 \pm \theta$, $90 \pm \theta$
d) $-\theta$, $360 \pm \theta$

8. The value of $\sin 180$ is ....
a) ∞
b) -1
c) 0
d) 1

9. The value of $\cos 150$ is ....
a) ∞
b) -1
c) 0
d) $-\sqrt{\dfrac{3}{2}}$

10. The value of $\sec 360$ is ....
a) ∞
b) -1
c) 0
d) 1

Answers : E.   1. b   2. d   3. a   4. b   5. d   6. b   7. d   8. c   9. d   10. d

F. 1. Which one of the following is **equal to $\sin \theta$**?

a) $\sin(90 - \theta)$
b) $\sin(180 - \theta)$
c) $\sin(270 + \theta)$
d) $\sin(-\theta)$

2. Which of the following **is/are equal to $-\sin \theta$**?

a) $\sin(360 - \theta)$
b) $\sin(180 + \theta)$
c) $\sin(270 + \theta)$
d) $\sin(-\theta)$

3. Which of the following **is/are equal to $-\tan \theta$**?

a) $\tan(360 - \theta)$
b) $\tan(180 - \theta)$
c) $\cot(270 - \theta)$
d) $\cot(90 - \theta)$

4. Which one of the following **is equal to $\cos \theta$**?

a) $\tan(360 + \theta)$
b) $\cos(180 - \theta)$
c) $\cos(270 - \theta)$
d) $\sin(90 - \theta)$

5. Which of the following **is/are not equal to $\csc \theta$**?

a) $\tan(360 + \theta)$
b) $\csc(180 - \theta)$
c) $\sec(270 - \theta)$
d) $\csc(90 - \theta)$

6. Which one of the following **is equal to $\cot \theta$**?

a) $\tan(90 - \theta)$
b) $\cot(180 - \theta)$
c) $\cot(270 - \theta)$
d) $\tan(90 + \theta)$

7. Which of the following **is not equal to** $-\sec\theta$?

a) $sec(-\theta)$ b) $sec(180 - \theta)$ c) $cot\,(270 - \theta)$ d) $tan\,(90 + \theta)$

8. Which one of the following **is equal to** $-\cot\theta$?

a) $tan(90 - \theta)$ b) $cot(180 - \theta)$ c) $tan\,(270 - \theta)$ d) $tan\,(-\theta)$

9. Which of the following **is/are equal to** $\sec\theta$?

a) $sec(-\theta)$ b) $sec(180 + \theta)$ c) $cot\,(270 - \theta)$ d) $cosec\,(90 + \theta)$

10. Which of the following **is/are equal to** $\tan\theta$?

a) $tan(360 + \theta)$ b) $tan(180 - \theta)$ c) $cot\,(270 - \theta)$ d) $cot\,(90 - \theta)$

Answers : F.   1. b   2. a, b, d   3. a, b   4. d   5. a, c, d

              6. a   7. a, c, d   8. b   9. a, d   10. a, c, d

G. 1. The domain of $\sin^{-1} x$ is

a) $\mathbf{R} - \{x : x = (2n + 1)\frac{\pi}{2}, n \in \mathbf{Z}\}$ b) $\mathbf{R}$

c) $\mathbf{R} - (-1, 1)$ d) $[-1, 1]$

2. The domain of $\cos^{-1} x$ is

a) $[-1, 1]$ b) $\mathbf{R}$

c) $\mathbf{R} - \{x : x = n\pi, n \in \mathbf{Z}\}$ d) $\mathbf{R} - (-1, 1)$

3. The domain of $\tan^{-1} x$ is

a) $[-1, 1]$ b) $\mathbf{R}$

c) $\mathbf{R} - \{x : x = n\pi, n \in \mathbf{Z}\}$ d) $\mathbf{R} - (-1, 1)$

4. The domain of $\cot^{-1} x$ is

a) $\mathbf{R}$ b) $\mathbf{R} - [-1, 1]$

c) $\mathbf{R} - \{x : x = n\pi, n \in \mathbf{Z}\}$ d) $\mathbf{R} - (-1, 1)$

5. The domain of $\operatorname{cosec}^{-1} x$ is

a) $\mathbf{R} - \{x : x = (2n + 1)\frac{\pi}{2}, n \in \mathbf{Z}\}$ b) $\mathbf{R}$

c) $R - [-1, 1]$
d) $R - (-1, 1)$

6. The domain of $\sec^{-1} x$ is

a) $R - (-1, 1)$
b) R
c) $R - \{x : x = n\pi, n \in Z\}$
d) $R - [-1, 1]$

Answers : G. 1. d 2. a 3. b 4. a 5. d 6. a

H. 1. The range (Principal value) of $\sin^{-1} x$ is

a) $[-\frac{\pi}{2}, \frac{\pi}{2}] - \{0\}$
b) $[-\frac{\pi}{2}, \frac{\pi}{2}]$
c) $[0, \pi] - \{\frac{\pi}{2}\}$
d) $(-\frac{\pi}{2}, \frac{\pi}{2})$

2. The range (Principal value) of $\cos^{-1} x$ is

a) $[-\frac{\pi}{2}, \frac{\pi}{2}] - \{0\}$
b) $[-\frac{\pi}{2}, \frac{\pi}{2}]$
c) $[0, \pi] - \{\frac{\pi}{2}\}$
d) $[0, \pi]$

3. The range (Principal value) of $\tan^{-1} x$ is

a) $[0, \pi] - \{\frac{\pi}{2}\}$
b) $(-\frac{\pi}{2}, \frac{\pi}{2})$
c) $(0, \pi)$
d) $[-\frac{\pi}{2}, \frac{\pi}{2}] - \{0\}$

4. The range (Principal value) of $\sec^{-1} x$ is

a) $[-\frac{\pi}{2}, \frac{\pi}{2}] - \{0\}$
b) $[-\frac{\pi}{2}, \frac{\pi}{2}]$
c) $[0, \pi] - \{\frac{\pi}{2}\}$
d) $[0, \pi]$

5. The range (Principal value) of $\text{cosec}^{-1} x$ is

a) $[-\frac{\pi}{2}, \frac{\pi}{2}] - \{0\}$
b) $[-\frac{\pi}{2}, \frac{\pi}{2}]$
c) $[0, \pi] - \{\frac{\pi}{2}\}$
d) $[0, \pi]$

6. The range (Principal value) of $\cot^{-1} x$ is

a) $(0, \pi)$ b) $[-\frac{\pi}{2}, \frac{\pi}{2}]$ c) $[0, \pi] - \{\frac{\pi}{2}\}$ d) $[0, \pi]$

Answers : H. 1. b   2. d   3. b   4. c   5. a   6. a

I. Miscellaneous questions

1. When the line OP is rotated from the initial line OX in the clock-wise direction through an angle θ (< 90), the value of θ is …………… *+ ve / – ve*

2. Neither $\sin \theta$ nor $\cos \theta$ can be numerically greater than unity. *True / False*

3. Both $\csc \theta$ and $\sec \theta$ cannot be numerically less than unity. *True / False*

4. If $\sin \theta = s$ then $\cot \theta = \frac{\sqrt{1-s^2}}{s}$. *True / False*

5. $\tan \theta = t \Rightarrow \cos \theta = \frac{\sqrt{1+t^2}}{t}$. *True / False*

6. **Two angles are said to be complementary when the sum of the two angles is 90^0. The complementary angle of θ is $90 - \theta$.**

   The sine of an angle θ is equal to the cosine of its complementary angle.

   True / False

7. **Two angles are said to be supplementary when the sum of the two angles is 180^0. The supplementary angle of θ is $180 - \theta$.**

   The supplement of 150 is 30. What is the supplement of $- 150$?

a) -150 b) 330   c) -330 d) 30

8. Which of the following formula changes sum/difference of two ratios into product of two ratios?

a) $\sin(A + B) = \sin A \cos B + \cos A \sin B$

b) $2 \cos A \sin B = \sin(A + B) - \sin(A - B)$

c) $\sin(A + B) \sin(A - B) = \sin^2 A - \sin^2 B$

d) $\cos C + \cos D = 2 \cos \dfrac{C+D}{2} \cos \dfrac{C-D}{2}$

9. Which of the following formula changes product of two ratios into sum of two ratios?

a) $\sin(A + B) = \sin A \cos B + \cos A \sin B$

b) $2 \sin A \cos B = \sin(A + B) + \sin(A - B)$

c) $\sin(A + B) \sin(A - B) = \sin^2 A - \sin^2 B$

d) $\cos C + \cos D = 2 \cos \dfrac{C+D}{2} \cos \dfrac{C-D}{2}$

10. $\cot(A - B) = $ ....

a) $\dfrac{\cot A \cot B + 1}{\cot B + \cot A}$
b) $\dfrac{\cot A \cot B + 1}{\cot B - \cot A}$
c) $\dfrac{\cot A \cot B - 1}{\cot B - \cot A}$
d) $\dfrac{\cot A \cot B - 1}{\cot B + \cot A}$

11. For a given value of θ, the value of $\sin \theta$ is unique. ***True / False***

The converse of the above, ie. for a given value of $\sin \theta$, θ need not be unique. ***True / False***

12. $\sin^{-1} x = \dfrac{1}{\sin x}$. ***True / False***

13. In the first quadrant,

i) sine .... (increases/decreases) from .... to ....

ii) cosine .... (increases/decreases) from .... to ....

iii) tangent .... (increases/decreases) from .... to ....

iv) cosecant ....(increases/decreases) from .... to ....

v) secant .... (increases/decreases) from .... to ....

vi) cotangent .... (increases/decreases) from .... to ....

14. In the second quadrant,

i) sine .... (increases/decreases) from .... to ....

ii) cosine .... (increases/decreases) from .... to ....

iii) tangent .... (increases/decreases) from .... to ....

iv) cosecant ....(increases/decreases) from .... to ....

v) secant .... (increases/decreases) from .... to ....

vi) cotangent .... (increases/decreases) from .... to ....

15. In the third quadrant,

i) sine .... (increases/decreases) from .... to ....

ii) cosine .... (increases/decreases) from .... to ....

iii) tangent .... (increases/decreases) from .... to ....

iv) cosecant ....(increases/decreases) from .... to ....

v) secant .... (increases/decreases) from .... to ....

vi) cotangent .... (increases/decreases) from .... to ....

16. In the fourth quadrant,

i) sine .... (increases/decreases) from .... to ....

ii) cosine .... (increases/decreases) from .... to ....

iii) tangent .... (increases/decreases) from .... to ....

iv) cosecant ....(increases/decreases) from .... to ....

v) secant .... (increases/decreases) from .... to ....

vi) cotangent .... (increases/decreases) from .... to ....

Answers I.

1. – ve

2. True. Reason : $sin^2\theta$ and $cos^2\theta$ are both positive.

If the numerical value of either $sin\ \theta\ or\ cos\ \theta$ is greater than 1 and

$sin^2\theta + cos^2\theta = 1 \Rightarrow$ either $cos^2\theta$ or $sin^2\theta$ is negative, which is impossible.

3. True. Reason : As $cosec\,\theta = \dfrac{1}{sin\,\theta}$, $sec\,\theta = \dfrac{1}{cos\,\theta}$ and the above result.

4. True   5. False   6. True   7. b

8. d   9. b   10. b   11. True, True

12. False, since $sin^{-1}x \neq (sin\,x)^{-1} = \dfrac{1}{sin\,x}$

13. In the first quadrant,

  i) sine increases from 0 to 1   iv) cosecant decreases from ∞ to 1

  ii) cosine decreases from 1 to 0   v) secant increases from 1 to ∞

  iii) tangent increases from 0 to ∞   vi) cotangent decreases from ∞ to 0

14. In the second quadrant,

  i) sine decreases from 1 to 0   iv) cosecant increases from 1 to ∞

  ii) cosine decreases from 0 to −1   v) secant increases from − ∞ to −1

  iii) tangent increases from −∞ to 0   vi) cotangent decreases from 0 to −∞

15. In the third quadrant,

  i) sine decreases from 0 to −1   iv) cosecant increases from −∞ to −1

  ii) cosine increases from −1 to 0   v) secant decreases from − 1 to −∞

  iii) tangent increases from 0 to ∞   vi) cotangent decreases from ∞ to 0

16. In the fourth quadrant,

  i) sine increases from −1 to 0   iv) cosecant decreases from −1 to −∞

  ii) cosine increases from 0 to 1   v) secant decreases from ∞ to 1

  iii) tangent increases from −∞ to 0   vi) cotangent decreases from 0 to −∞

Available soon

PROF. MSDOSS MATHS BOOK SERIES III

MATHEMATICS

CALCULUS II

INTEGRAL CALCULUS (Volume I)

FORMULAE PRACTICE WORKBOOK

FOR THOSE STUDENTS WHO WANT TO HAVE A STRONG BASE IN INTEGRAL CALCULUS FORMULAE & VARIOUS METHODS

www.ingramcontent.com/pod-product-compliance
Lightning Source LLC
Chambersburg PA
CBHW080716190526
45169CB00006B/2400